The Illustrated Guide to

CRYSTALS

The Illustrated Guide to

CRYSTALS

JUDY HALL

The information given in this book is not intended to act as a substitute for medical treatment, nor can it be used for diagnosis. Crystals are powerful and are open to misunderstanding or abuse. If you are in any doubt about their use, a qualified practitioner should be consulted, especially in the crystal healing field.

ACKNOWLEDGMENTS

The staff at Earthworks, especially Steve and Jacquie, gave freely and patiently of their time and expertise. They supplied many of the crystals in this book. For this, and much, much more, love and thanks. My thanks also to Pat Goodenough, Barbara Reeve, Trudi Green, and Jan Ross for introducing me to many of my crystal friends.

First published in Great Britain in 2000
by Godsfield Press Ltd.
Laurel House, Station Approach,
Alresford, Hampshire SO24 9JH, U.K.
www.godsfieldpress.com

10 9 8 7 6

Designed for Godsfield Press by
The Bridgewater Book Company

ISBN 1-84181-006-1

Photographer Guy Ryecart
Illustrations by Nicola Evans (*78–85*),
Sharon Harmer (*10, 15, 93*),
Andrew Kulman (*29, 33, 41, 49*),
Catherine McIntyre (*6, 20, 26, 46, 56, 60, 62, 76, 86*)
Diagrams by Kate Nardoni of MTG (*31, 78–85*)

For further acknowledgments and picture credits *see page 128*.

Printed and bound in China

Contents

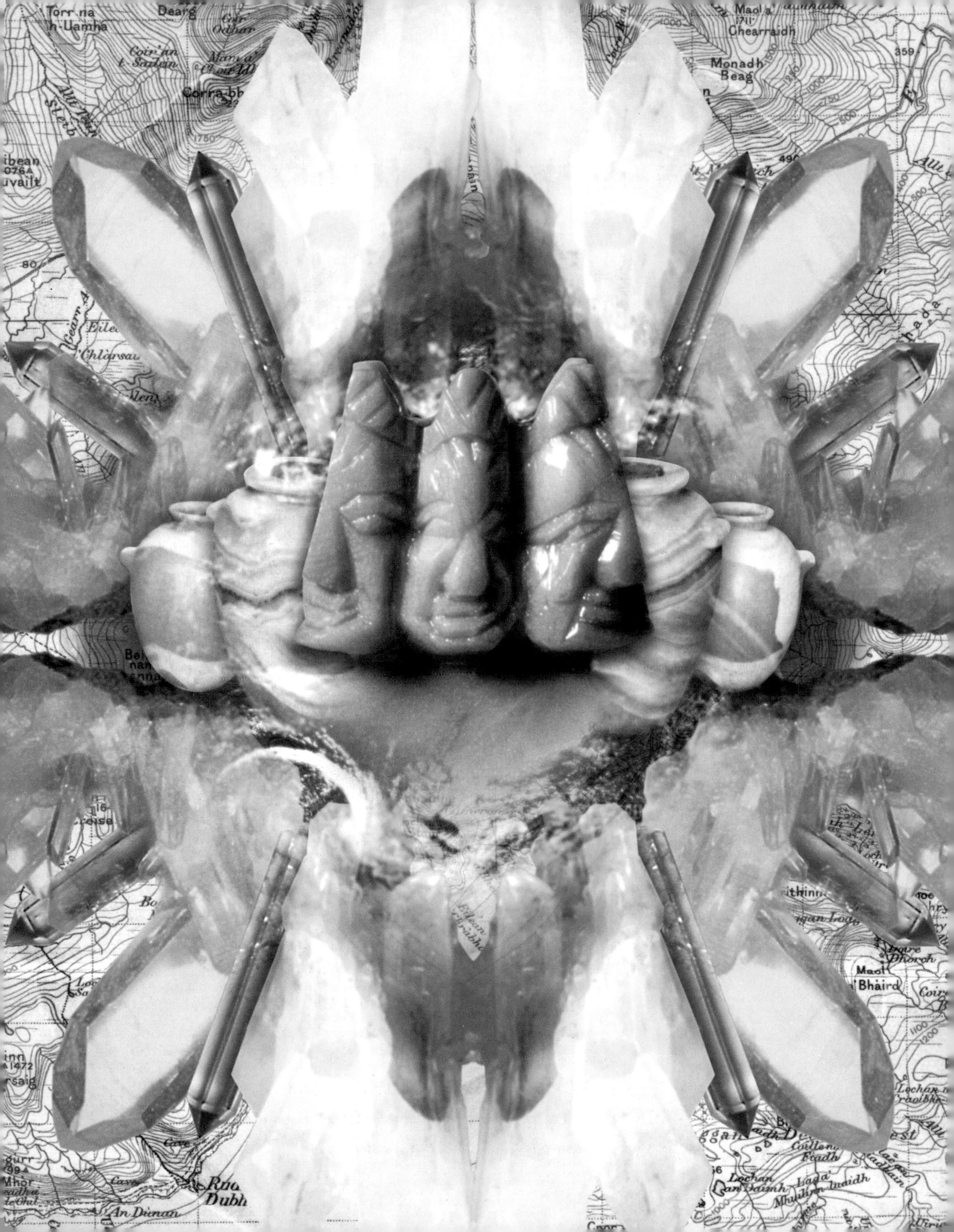

Torr na
h-Uamha
Chearraidh
Monadh
Beag
1750
Maol
Bhaird
Cave
An Dunan
Lochan

1 Living with Crystals

For thousands of years, crystals have been used for healing, for protection, and for adornment. Amber was probably the first crystal used for decoration. It is found in Stone Age deposits, as is Jet. Neolithic people were accompanied to the Next World by polished crystal mirrors (for seeing the future) and gemstone beads. To ancient peoples, crystals were sacred gifts from Mother Earth. As time went on, they retained their magical and spiritual qualities. Priests, medicine men, rulers, and shamans wore the powerful gems as symbols of spiritual and temporal authority.

In the Biblical saga Exodus, a breastplate is made, at the instigation of God, from 12 gemstones. Worn over the heart, it would protect the High Priest through its immensely strong spiritual powers.

Egypt's rulers were buried with gems to conduct their souls to the Other World: amulets and talismans of Lapis Lazuli, Turquoise, Carnelian, and Quartz. The living, too, had their amulets of power, their protective talismans, their love charms.

In India in 400BCE, a Sanskrit document laid down the astrological properties of gems. Since that time, Indian astrologers have advised those struck with misfortune on which gem to wear to avert the problem.

Throughout the ancient world, the Americas, the West, and the East, the richness, rarity, and variety of gemstones were valued. Some gems traveled surprisingly far from their source as ancient trade routes carried gems all over the world. Many were ground into medicines, and remedies from as far apart as Egypt, South America, and China contained precious gem substances. Crystals have long been considered objects of great beauty and power, and we still consider them so today.

Crystals, such as Carnelian, Jet, and Lapis Lazuli, have been used since the start of human history to adorn, heal, protect, and avert misfortune.

Crystals in the modern world

Crystals have never lost their fascination, or their value. Humankind still faces enormous danger in order to pluck crystals from deep within Mother Earth. Their unique beauty, versatility, and intrinsic qualities mean that in the modern world they are used for decoration, in industry, in healing, and in medicine. Precious gems command high prices, but less flamboyant crystals can have a powerful effect, especially in the rapidly expanding field of crystal healing.

RIGHT *This elaborate crown comes from the British Crown Jewels. English kings and queens have been crowned with jewel-encrusted symbols of authority for at least a thousand years.*

BELOW *Crystals can be used for adornment – because of their intrinsic beauty – but they can also be selected because of the special qualities of the stone.*

CRYSTAL JEWELS

People everywhere adorn themselves with beautiful jewels. They may not know the properties of the stones, but they instinctively feel good when wearing them. Everyone has a favorite piece, a gift from a lover perhaps, or something treasured from childhood or inherited. Jewelry crosses cultural boundaries and represents personal wealth. Tribal women in their dowry jewels are soul-sisters to Western women with their diamonds.

CRYSTAL LOVE

Crystals signify fidelity. Exchanged by lovers to symbolize their union, they have sentimental value. The stones have a language of their own: some mean faithfulness; others pure love. When one chooses a stone to symbolize the qualities that a relationship needs, often times the stone brings those qualities into the relationship. Crystals can be used to revitalize love, attract a soulmate, or heal a broken heart.

CRYSTAL AUTHORITY

Crystals have always been used to demonstrate authority. They are symbols of power, indicating rank and privilege. Although the meaning behind the crystals has, to a large extent, been lost, countries still have Crown Jewels or insignia that not only symbolize majesty but also transfer the power of the jewels onto the wearer.

CRYSTAL INDUSTRY

Crystals have a more prosaic application. They can be found throughout industry, storing and focusing energy. They cut and transmit, absorb and regulate. Diamond-bit drills are highly effective cutting tools. Quartz crystals power our watches, control computers, emit radio waves. As part of state-of-the-art technology, crystals go into space on the "shuttle." Such crystals are manufactured and carry little of the true power of a natural crystal, but the modern world could not function without them.

ABOVE *Manufactured Quartz crystals guide the navigational systems and communication equipment of the most sophisticated technology in our world.*

CRYSTAL HEALING

The ancient art of crystal healing uses the gentle, transformative properties of gems. Specific resonances of crystals transmit energy. Such crystals can be used at a distance or placed on the body. They dissolve stress, remove blockages, support new intentions, and bring harmony into the environment. They neutralize negative energies, draw energy away from an overstimulated area, or reenergize a depleted one. Their effect is pleasant: they induce a feeling of well-being and harmony.

But it is not only in complementary therapies that crystals are used. Modern conventional medicine has rubies in its surgical lasers, used for treatment of gallstones, for example. Silicon chips – manufactured crystals – are a vital part of pacemakers and other life-saving equipment.

Natural medicine, too, uses crystals. Gem elixirs work by subtle energy transference. This approach is not new. Elixirs have been used for thousands of years, just as gems have long been ground and included in potions and pills, a practice that continues today in many forms of traditional medicine.

BELOW *Many forms of traditional medicine still use finely ground crystals in their remedies – as they have done for thousands of years. Gem remedies are also used in crystal healing.*

What crystals do

Crystals transmit and receive energies. They transmute and transform, attract or repel. The most powerful may appear dull – until the beneficial effects are felt. They bring serenity to a home, peace to an individual. They protect or heal, attract abundance, and create love. Objects of beauty in their own right, they enhance the environment wherever they are placed.

ABOVE *Crystals enhance the natural energy field of the human body. They can heal, repair, and energize – the effect feels almost magical.*

CRYSTAL ENERGY

It is no wonder that ancient people were so much in awe of crystals. They have an almost magical effect. Crystals are used to treat food, water, plants, animals, and human beings. They can even help your car run more efficiently and use less gas! They enhance meditation, deepening attunement with universal energies; aid manifestation, drawing close whatever is necessary; and offer protection, absorbing negative energies and strengthening positive ones.

Crystals magnetically attract certain things, or repel unwanted energies. They resonate at different frequencies and can store, or retrieve, information.

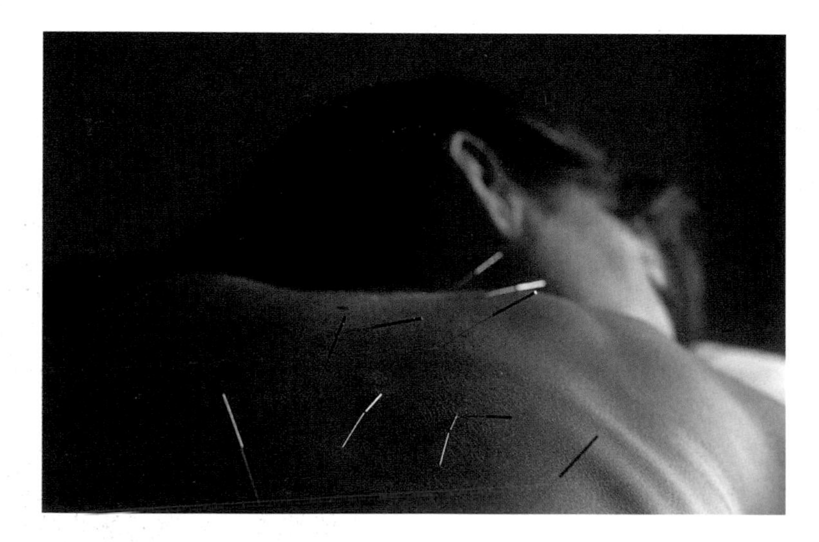

BELOW *When the needles are coated in Quartz, the effect of acupuncture is greatly enhanced.*

ENERGY TRANSMITTERS

The crystalline structure of crystals means they can absorb, focus, and then transmit subtle electromagnetic energy, a quality that is used in crystal healing. No one is quite sure exactly how crystals communicate their energies to us. Silicon is an important component of both crystals and our bodies, and energy might be transferred between the two. Most people believe that crystals work through the aura, or etheric body, the subtle magnetic field that surrounds our physical body.

Science has shown that if a crystal is placed in an energy field, it collects that energy and contains it. It might change or transmute the energy in the process. Some crystals, such as Quartz, amplify energy. The crystal then radiates energy out again. The exception is black crystals, which absorb energy but do not release it, making them extremely useful as protection against unpleasant or hostile vibrations.

BLUE QUARTZ

If an acupuncture needle is coated in Quartz, its effect is enhanced by as much as 10 or 12 percent. Simply holding a Quartz crystal at least doubles the aura, an effect that can be photographed by a kirlian camera.

HEALING ENERGIES

The ability of crystals to focus energy means that they can be used for specific tasks, such as directing healing energy to a point on the body or to an emotional blockage. The dis-ease is gently dissolved, and any imbalance is corrected. A wide variety of conditions respond to crystal healing. Crystals not only heal people, they also heal animals and the environment. They can be programmed to radiate good vibrations out into the surrounding area or to absorb electromagnetic stress.

COSMIC TRANSCEIVERS

This ability to transmit energy has been used for thousands of years. Birthstones, for instance, receive cosmic energy, focus it, and then resonate it to the wearer. In effect, they act as a transceiver for cosmic energies, linking the earth and the sky. This aligns the wearer with cosmic forces that flow throughout the universe, enhancing health and well-being. Such stones never lose their effectiveness; they cannot wear out. They may need cleansing and re-energizing from time to time, but they retain their power.

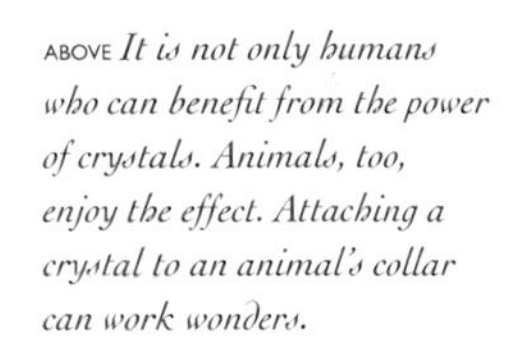

ABOVE *It is not only humans who can benefit from the power of crystals. Animals, too, enjoy the effect. Attaching a crystal to an animal's collar can work wonders.*

LEFT *A carefully placed crystal is not only a decorative feature but can improve the energy in a room.*

QUARTZ CLUSTER

ROSE QUARTZ

CRYSTAL POINT

SMOKY QUARTZ

Choosing crystals

Crystals attract or repel. If a crystal is meant for you, it will find its way to you. When its benefit is over, it will pass to someone else. The best crystal chooses you. It might be a gift or one that catches your eye in the crystal shop, but it will resonate with your deepest being. It will become like a much-loved friend. Remember that the showiest crystal may not be the most effective for you, and try to take your time in choosing one.

CRYSTAL ATTRACTION

Crystals grab your attention. You suddenly find your eye drawn to one and it will not let go: you simply have to have it. Or you idly dip your hand into a tub of stones and one sticks to your fingers. These are the crystals for you, and they are letting you know it.

Choosing crystals works for your friends and family, too. You see a crystal, and a person comes to mind. You feel that it would make a beautiful gift. Follow that instinct. You can picture a person and ask to be shown the right crystal. It will be beneficial.

TUNING IN

Crystals transmit energy fields. If you hold them, they communicate with you. Their energy feels good – or not, as the case may be. When choosing a crystal, do so when you have time to spare, when your energy is quiet and calm and you can sit with the crystals a while before you make your choice. As you choose, bear in mind that biggest is not necessarily best, nor is flamboyant beauty or high cost always a sign of healing power. Some of the quietest stones have the greatest effect.

ABOVE *Although crystals may look alike, each one has its own unique character. In a bowl of crystals, there will be one that is beautifully attuned to your energies, and it will find you.*

RIGHT *Take your time when choosing a crystal. Hold it, and tune in to its energy. It will soon let you know if it is the one for you.*

To tune in, hold a crystal in your hands. Look at it. Feel its texture, its weight. Let the color and shape speak to you. Close your eyes and allow the crystal to whisper into your inner ear. If you feel good at the end of this process, then this is the crystal for you. If you do not feel so good, then it is worth looking at the properties of the crystal to see whether it is bringing up something that you are not yet ready to hear or something you are avoiding. Come back to it later.

Another way to attune to a crystal is to go into a crystal shop when you feel needy or vulnerable. You will be drawn to a particular color. Put your hands out and feel the stones. The right stone will feel good; it will bring comfort to you. This is the stone that will transfer its healing properties to you.

MOVING ON

One day you might find that it is time for your crystal to move on. Crystals have distinct personalities. They are like beloved friends with strong minds of their own. They do their work, stay a while, and then move on. You may lose one, or suddenly have an urge to pass it on. It is the nature of crystals that they do move on. Allow them to go freely. A new one will soon find its way to you.

Before passing a crystal on, remember to cleanse it and to show your friend how to program it for his or her special needs.

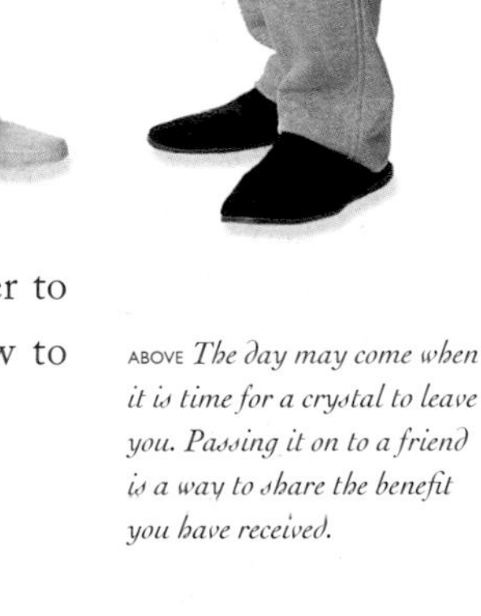

ABOVE *The day may come when it is time for a crystal to leave you. Passing it on to a friend is a way to share the benefit you have received.*

Cleansing crystals

Crystals absorb and emit energy. This means that crystals absorb negative energies during mining or transportation, while awaiting sale, or through use. New-to-you crystals need to be purified, and all crystals should be cleansed regularly to ensure optimum energy.

BELOW *Smoky Quartz is particularly good at absorbing negative energies. This means that it needs to be cleansed regularly, especially if used for crystal healing.*

CRYSTAL CLEAN

Crystals absorb energy from their surroundings. If you purchase a crystal, it will have the energy imprint of everyone who has handled it. Crystals in your home absorb energetic vibrations, as do those worn for decoration or protection. Some crystals are used to draw off energy, as in healing or psychic protection. Others are programmed to change the vibration, as in attracting a soulmate or abundance. Crystals may absorb negative energy so that positive can be released. If a crystal has picked up unpleasant energy, it can easily pass it on. Crystals need to be purified regularly: daily or after each use if facilitating healing, and weekly or monthly if they have a decorative function.

A PURIFICATION RITUAL

Purification is part of the ritual of using crystals and there are many ways of doing this, some of which take several hours, others only a minute or two. Certain methods are more suitable for particular crystals than others. For instance, soft or friable stones should not be immersed in water, and care must be taken not to scratch polished crystals.

As you cleanse your crystal, hold the intention in your mind that any negative energies will be transmuted into positive ones and that the crystal will be reenergized.

Natural forces, such as the sun, moon, water, or plants, are effective purifiers of negative or stale energies. Choose the most appropriate method for your crystal, according to its type and the tools you have at hand.

CLEANSING WITH WATER

SALT WATER

Most crystals enjoy being immersed in the ocean for an hour or so (place small ones in a mesh bag to avoid loss), after which they can be rinsed in pure water and allowed to dry naturally in the sun to reenergize. A handful of sea salt dissolved in water works just as well.

FLOWING WATER

Holding the crystal under a waterfall, in a stream, or beneath the kitchen faucet will rapidly cleanse the energies. Allow to dry naturally in the sun.

RIGHT *Atomized spray can be used to cleanse crystals; alternatively, a few drops of purpose-made cleanser added to water instantly clears the crystal.*

ABOVE *Certain crystals resonate with the energy of the sun – especially yellow, orange, and red ones. Others, such as white stones, appreciate the more gentle energies of the moon.*

Smudging

Hold the crystal in smoke from a sage, cedar, or sweetgrass smudge stick or use incense. A feather can be used to fan the smoke gently across the crystal.

Quartz cluster

Stand the crystal on a large cluster of Quartz or other energizing crystal for approximately 12 hours. Stones that have a natural affinity with the moon can be left in moonlight, while those that empathize with the sun will benefit from sunlight.

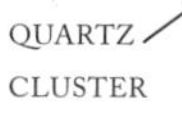

RIGHT *Quartz cleanses other crystals effectively. Here Clear Quartz has been used, but a Smoky Quartz or Amethyst cluster is also appropriate.*

Visualization

Visualize a column of pure white or golden light coming down to surround the crystal, cleansing and transmuting the energies. Hold that thought until the crystal energy shines out once more.

Crystal Clear

By far the easiest method of cleansing crystals is with Crystal Clear or other special cleanser. A few drops of Crystal Clear in the washing water or sprayed from an atomizer will clear any negativity and restore the energy of the crystal. (Atomized spray cannot hurt delicate or easily crumbled crystals.)

Salt

Delicate or friable stones that would be harmed by water can be placed directly into salt or rock-salt crystals and left there for a few hours to cleanse themselves.

Programming crystals

To make a crystal your own, you need to program it. Programming attunes the crystal's energies to your own frequency and desires. This ensures that whatever you wish from your crystal will come about. Before programming a crystal, it is usual to dedicate its energies to the highest good of all.

CHRYSOPRASE

DEDICATING A CRYSTAL

When you receive a new crystal, it can be dedicated immediately after cleansing. Dedication is simple, but makes an enormous difference to how effective the crystal will be. Dedication focuses the intention that the crystal will be used positively for the highest good of all concerned.

Hold the crystal in your hands, picture light surrounding it, and state clearly that the crystal will only be used for the highest good, in light and love.

When you have finished the dedication, you can either move directly on to programming the crystal or you can leave the crystal in the light of the sun or moon to energize.

Crystal Directory

GREEN CRYSTALS

GREEN APOPHYLLITE
PROPERTIES Activates heart chakra, promotes a forthright heart. Absorbs universal energy. Aids fire walking.
See Apophyllite, page 99, for general information.
POSITION Place on the third eye.

GREEN KUNZITE
PROPERTIES Helps to ground spiritual love.
See Kunzite, page 104, for general information.
POSITION Hold or place on the appropriate point.

GREEN-PINK CRYSTALS

UNAKITE
SOURCE South Africa
APPEARANCE Mottled
PROPERTIES A "stone of vision," balances emotions with spirituality. Facilitates rebirthing, integrates information from past that create blockages, gently releases conditions inhibiting growth. Reaches root cause of dis-ease. Treats reproductive system, creates weight gain, aids healthy pregnancy.
POSITION Hold or place on the appropriate point.

WATERMELON TOURMALINE
APPEARANCE Pink enfolded in green
PROPERTIES Aids understanding of situations. "Super-activator" of heart chakra, linking it to higher self. Treats emotional dysfunction. Resolves conflict.
See Tourmaline, page 117, for general information.
POSITION Hold or place on the appropriate point.

GREEN AQUAMARINE
See Aquamarine, page 110.
GREEN IRON PYRITE
See Iron Pyrite, page 108.
GREEN CHRYSOCOLLA
See Chrysocolla, page 112.
GREEN TOPAZ
See Topaz, page 97.

GREEN SERPENTINE
See Serpentine, page 124.
GREEN MUSCOVITE
See Muscovite, page 105.
GREEN LEPIDOLITE
See Lepidolite, page 121.
GREEN CAT'S EYE
See Cat's Eye, page 106.

GREEN DIOPTASE
See Dioptase, page 113.
GREEN ONYX
See Onyx, page 124.
GREEN SELENITE
See Selenite, page 99.

118

ROYAL-INDIGO CRYST

SAPPHIRE
SOURCE Burma, Czech Republic
India
APPEARANCE Bright, transpare
PROPERTIES Relaxes, focuses,
releasing unwanted thoughts
promoting peace of mind an
physical, mental, and spiritu
balance within body. Suppo
purpose. Releases depressio
confusion, aids concentrati

AZURITE
SOURCE USA, Australia,
Namibia, Russia
APPEARANCE Very smal
when tumbled)
PROPERTIES Stimulate
psychic development

LAPIS LAZU
SOURCE Russia, A
Egypt
APPEARANCE Op
PROPERTIES A "s
enlightenment,
psychic abilitie
deep peace. Po
Stimulates hi
objectivity an

SODAL
SOURCE N
Greenland

ABOVE *The Crystal Directory, pages 94–125, will help you select exactly the right crystal for your purpose.*

TIGER'S EYE

ABOVE *If you are going to wear or carry a crystal, you need first to program it so that it fulfills its purpose. Remember, too, to cleanse it regularly.*

PROGRAMMING

Crystals can be programmed for many purposes. You may want protection, prosperity, peace, or love. You may be looking for healing, or intending to use the crystal to heal someone else. (If so, permission should be sought before programming begins.) You may be improving your home or work environment, or seeking angelic contact. It is important that programming is well thought out and properly focused. Imprecise wording confuses crystal energies, but meticulous wording directs energy precisely.

Select a crystal that resonates with the intention you have for it. (The Crystal Directory at the end of the book will help you here.) If you are looking for calmness and peace, a stone to meditate with or to let go of stress, one of the calming stones from the green color range could be appropriate. If you want an energizing stone, one that brings positive results quickly, stimulating things and getting them moving, a crystal from the red range could be a good choice. If you feel red is too forceful, then a yellow or orange crystal is more mellow.

When you have precisely formulated your intention, hold the crystal in your hands. Sit quietly with it, letting yourself attune to its energy. Let yourself be open to any higher guidance that may be with you. When you feel totally in harmony with the stone, state your intention firmly and clearly. (You may repeat this several times to anchor the intention into the crystal.) When you intuitively feel that the programming is complete, put the crystal down. Detach your energies from the crystal by taking your attention away from it.

When the crystal has been programmed, you may like to make a special time each day when you sit with it, asking that the energies be transferred to you. Depending on its purpose, you can wear the crystal, place it in your pocket, or put it under your pillow or by your bed at night.

PROGRAMMING A CRYSTAL FOR SOMEONE ELSE

It is far better for someone to program his or her own crystal. If this is not possible, simply ask that the crystal will work for the person's highest good. Try not to impose your own ideas onto the crystal.

ABOVE *Pendulums can be dedicated and programmed before use so that they will always speak the truth.*

RIGHT *When programming or dedicating a crystal, focus all your energy on it and state your intention clearly and precisely.*

Crystal beauty

Crystals take many forms. Some are square, others pointed; some round, others flat. There are granular stones and transparent crystals. Certain crystals are one color, others a combination. No matter how they look on the outside, inside they have a stable, orderly structure. A beautiful, repeating, three-dimensional lattice created from ancient atoms holds the unchanging crystalline form.

RAINBOW OBSIDIAN

RIGHT *Obsidian was formed through heat and pressure as molten volcanic material cooled below ground, giving a characteristic "glassy" look.*

BELOW *This enormous piece of Amber is as long as a man's forearm. Strictly speaking, it is not a crystal, because it was formed from solidified tree resin, but its beauty has been highly prized for eons.*

WHAT IS A CRYSTAL?

Crystals are born deep within the womb of Mother Earth. A crystal has been defined as a "substance solidified in a definite geometric form." Strictly speaking, crystals are large, faceted, and clear, like Amethyst, but any stone with a crystalline structure can be referred to as a crystal. When a crystal has clear facets, coming to a point, it is known as "a point."

This geometric form results from the way in which crystals are created. Millions of years ago, superheated gases and mineral solutions forced their way up out of the earth's molten interior toward the surface. As they cooled, atoms formed orderly patterns, three-dimensional lattices repeating throughout the crystal and holding the internal structure stable. Crystals belong to "families," such as Quartz. Stones from a family share basic qualities, however different they appear.

The appearance of a crystal is affected by the temperature and pressure at which it formed, how slowly it cooled, and the amount of space available for it to grow. Bubbles in the magma or a wider fissure created space for bigger crystals. Hard, transparent crystals such as Diamonds result from high pressure and temperature. Softer stones, such as Calcite, were made at lower temperatures. Certain crystals underwent further changes, melting and re-forming as conditions altered.

Not all crystals are translucent gems. Sometimes the full beauty does not emerge until the stone has been cut into facets to refract the light. Most gem-quality crystals also come in the form of more cloudy, less valuable stones that, nonetheless, work perfectly well for healing, meditation, and similar applications. Other crystals are grainy and look more like stones. There are a few crystals that do not conform to crystalline structure, such as Obsidian, formed from volcanic glass, and Amber, fossilized tree resin, but almost all crystals can be recognized by their own unique lattice.

Many of the small stones in crystal shops have been tumbled, a process that polishes the stone and alters its appearance but does not affect its properties.

The shape of a crystal affects how it transmits energy and modifies how that energy affects its surroundings. A single crystal has a different effect to that of a cluster or a geode, even when it is the same stone.

AMBER

CRYSTAL SHAPES

A LONG, POINTED CRYSTAL *focuses the energy in a straight line. It transmits energy or draws it off, depending which way it is pointing. Shaped crystal wands or large natural crystals are often used in healing.*

A DOUBLE-TERMINATED CRYSTAL *(a crystal having points at both ends) radiates or absorbs energy at its extremities. Such crystals balance and integrate spirit and matter. Double terminations break old patterns and are useful in treating addictions. They can also help to develop telepathy.*

GEODES, *rounded, cavelike stones, diffuse and contain the energy of the crystal.*

CRYSTAL CLUSTERS *radiate the energy of the crystal to the surrounding environment. They are particularly useful for cleansing energy in a room, or cleaning other crystals.*

EGG-SHAPED CRYSTALS *can be used over the body to detect and rebalance blockages. The pointed end can be used as a tool for therapies such as acupressure or reflexology.*

BALLS, NATURAL OR SHAPED, *emit energy in all directions equally. Used as "windows" they can transport energy from another time and place: past or future.*

SQUARE CRYSTALS *consolidate energy.*

PYRAMID-SHAPED CRYSTALS *tightly focus energy through the apex.*

2 Crystal Love

Crystals have been used for thousands of years to express love and stimulate passion. Love, be it gently romantic or powerfully erotic, radiates from many stones. Stones can attract a soulmate or put a zing into an existing relationship. They can change how you feel about yourself, and make you more open to love.

It is said that love makes the world go around – something with which crystals would wholeheartedly agree! You can feel love radiating out from heart chakra opening stones, such as Rose Quartz, Rhodochrosite, or Kunzite, which can be very soothing when worn over the heart. They have a gentle energy that helps you to accept and love yourself – the prerequisite to being loved by someone else.

If you are looking for something with more verve and eroticism, the passionate orange and red stones such as Carnelian or Garnet are for you. Scatter them around your bedroom and the aphrodisiac effects soon make themselves known! If the feelings are there, but not much action, Green Tourmaline will come to your aid. If sharing physical pleasure brings on your inhibitions, try Pink Tourmaline. It assures you that you can trust your body and let love in. Watermelon Tourmaline combines the two.

Rose Quartz is the perfect heart stone. It heals the heart and it attracts love.

One of the most enjoyable ways to experience crystal love is to take a bath with your favorite stone. Choose one that radiates out the qualities you seek: passion, romance, self-love, healing for your heart. Cleanse the crystal and place it in the water, to which you can add a few drops of rose oil. Take a long, hot soak and absorb the energies. If you share the bath with your partner, so much the better. At night, slip the stone under your pillow to reinforce its effects.

Attracting a soulmate

The search for a soulmate is universal. A soulmate is someone with whom you share not only love but also your deepest thoughts and feelings. A crystal can attract just such a soulmate into your life. Keep the appropriate crystal with you, or place it in your bed.

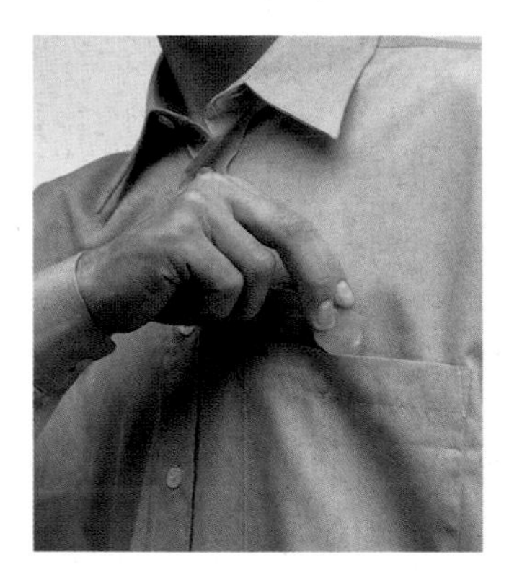

ABOVE *Put the right crystal into your pocket and it will attract a soulmate to you.*

ROSE QUARTZ

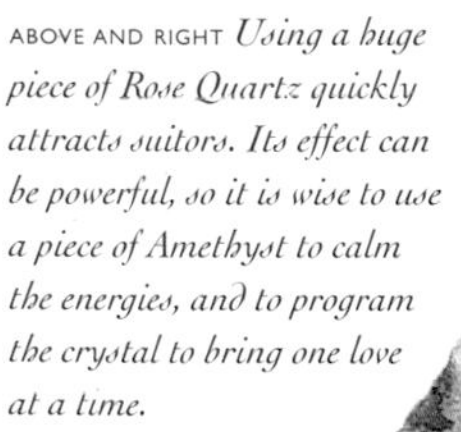

ABOVE AND RIGHT *Using a huge piece of Rose Quartz quickly attracts suitors. Its effect can be powerful, so it is wise to use a piece of Amethyst to calm the energies, and to program the crystal to bring one love at a time.*

ROSE QUARTZ

The stone of unconditional love, Rose Quartz has a translucent, soft pink sheen. It gently dissolves blockages to love. If you have not been able to love yourself, this sympathetic stone brings forgiveness and self-acceptance. When you love yourself, the way opens to attract a soulmate.

A large piece of Rose Quartz will quickly draw love into your life. Its effect is powerful, and you could find yourself overwhelmed with suitors! An Amethyst crystal placed alongside moderates the attraction.

AVENTURINE

Aventurine is another member of the Quartz family. Grainy in appearance, it is speckled with Hematite and Mica, dull on the surface yet sparkling in light. Green Aventurine activates, clears, and protects the heart, enhancing the ability to attract love and increasing empathy. It is a wonderful stone for mature love.

AMETHYST

MAGNETITE

MAGNETITE (LODESTONE)

Magnetite is a dark, granular-looking stone formed from iron. Because it is magnetic, it is the ideal stone for attracting love. In olden times, it was used to test the loyalty and fidelity of a wife. A man would place it beneath his wife's pillow and if she fell out of bed, she was no longer virtuous.

RHODOCHROSITE

A banded pink, yellowish-white, and orange stone, Rhodochrosite pulsates with love. Used in meditation it can link to a twin soul, a spiritual soulmate.

SOULMATE CRYSTAL

A soulmate crystal is two clear crystals of similar size, usually Quartz, joined together side by side. It is also known as a tantric twin.

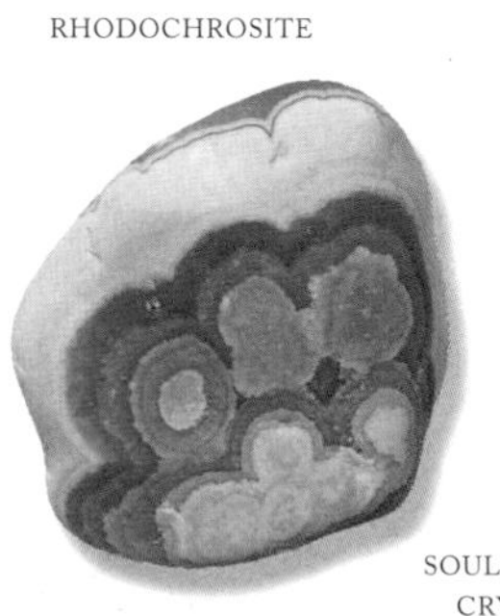

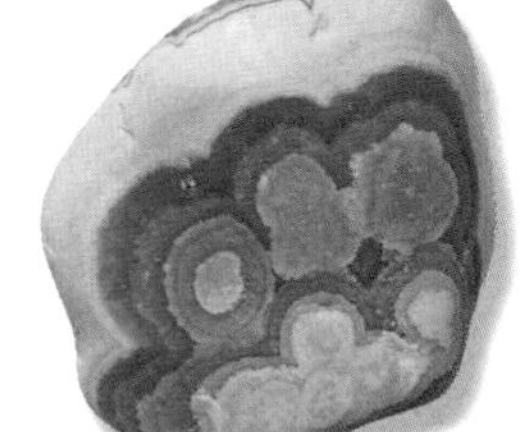

RHODOCHROSITE

SOULMATE CRYSTAL

Healing the heart

Wounds to the heart make you cautious about love. No matter how much you yearn for a relationship, you hold back. A broken heart is a painful experience, creating blockages, but crystals are perfect for gently dissolving such blockages and healing the wounds of love. Simply place the crystal on your heart and wait for it to do its work.

LAVENDER JADE

Jade is a soft, soapy stone that comes in two types: Jadeite, which is translucent; and Nephrite, which has a creamy consistency. Lavender Jade aids people who have been hurt by love, dissolving the pain and allowing them to touch the gentleness within themselves. This loving acceptance heals the heart.

OBSIDIAN

Obsidian is a hard, shiny, volcanic glass, with the velvet blackness of total darkness to help release old loves, cleansing the heart of outgrown ties and painful memories. Apache Tears, transparent, water-smoothed Black Obsidian pebbles, are "shed tears," bringing insights into pain and creating acceptance. Rainbow Obsidian, too, can cut the cords of old loves, gently releasing the hooks that others have left in the heart.

KUNZITE

Kunzite is a form of spodumene or lithium. Pink Kunzite activates the heart chakra, purifying it and filling it with peace. It removes emotional debris left behind from past relationships, making it the perfect stone for healing a broken heart.

AGATE

Agate is a member of the Chalcedony family. It is a banded stone with a somewhat waxy appearance and is often transparent at the center. It overcomes bitterness of the heart and eliminates inner anger. Agate fosters love, opening the way for positive relationships.

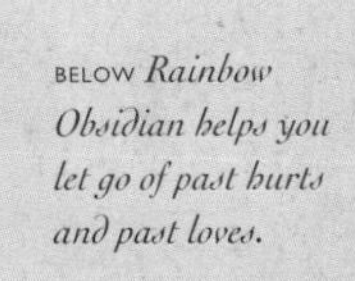

PINK AGATE

KUNZITE

LAVENDER JADE

BELOW *Rainbow Obsidian helps you let go of past hurts and past loves.*

Creating positive relationships

Crystals create positive relationships. They draw in qualities that are lacking, transmute negative factors, and nourish love. Crystals help you to work through conflict, and to transform a dysfunctional relationship into a healthy, happy one. For best effect, wear the stone over your heart and place it by the bed as well.

TURQUOISE

Ranging in color from sky-blue to green, Turquoise unites the earth and the sky. It is said that if Turquoise is offered as a pledge of friendship, the spirit dwelling in the stone will transfer its assistance to the recipient. In relationships, Turquoise's greatest contribution is helping to tune in to, and understand, other people.

GREEN TOURMALINE

Tourmaline has a long, vertically striated structure. Green Tourmaline translates your feelings into action, as does Kunzite. Green Tourmaline can also help you to recognize a problem, get to the bottom of it, and deal with it constructively.

OPAL

Beautiful and iridescent, Opal fosters love, passion, loyalty, and faithfulness. It is often used for engagement rings for this reason. An emotionally responsive stone, Opal brings stability to a relationship.

RUBY

A vivid red, transparent corundum, Ruby fosters romance and marriage, integrity, devotion, and passion. Its positive energy helps in all matters of love and it can increase virility.

SUGILITE (LUVULITE)

Deep purple and lightly banded, Sugilite is also known as Luvulite because of its affinity with love. It represents the perfection of spiritual love and has the capacity to foster forgiveness. In relationships, it can protect the inner self from the shocks and disappointments of the world and has the power to eliminate hostility and jealousy.

BELOW *Crystals can bring harmony into all types of relationships, enhancing love between partners, and between parents and child.*

TURQUOISE

OPAL

SUGILITE

GREEN TOURMALINE

AMETHYST GEODE

RED JASPER

WATERMELON TOURMALINE

SARDONYX

JASPER

Jasper is an opaque Chalcedony and may be patterned or banded. Because Jasper enhances energy, it can prolong sexual pleasure. Red Jasper has the useful attribute of bringing problems to light before they become a threat to happiness.

DIAMOND

Composed of pure carbon and the product of thousands of years of intense heat, Diamonds are one of the most precious gemstones. Traditionally used for engagement rings, Diamonds bond a relationship and enhance the love of a husband for his wife.

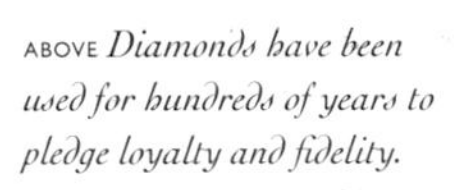

ABOVE *Diamonds have been used for hundreds of years to pledge loyalty and fidelity.*

SARDONYX

If it is lasting happiness that you seek, Sardonyx is the answer. This banded Onyx brings joy and commitment to marriage and live-in relationships.

GEODES

The womblike shape of a crystal geode, such as Amethyst, brings compatible friendship and harmony into a community.

WATERMELON TOURMALINE

Pale pink enfolded within green, Watermelon Tourmaline shows how opposites work together to bring about balance. Green Tourmaline is a problem solver, while Pink Tourmaline brings more love into a situation. Watermelon Tourmaline is helpful in resolving conflict.

OTHER CRYSTALS FOR POSITIVE RELATIONSHIPS

ROSE QUARTZ
Brings unconditional love into the relationship

GREEN JADE
Improves dysfunctional relationships

RHODOCHROSITE
Aids emotional expression

TOPAZ
Teaches how to love

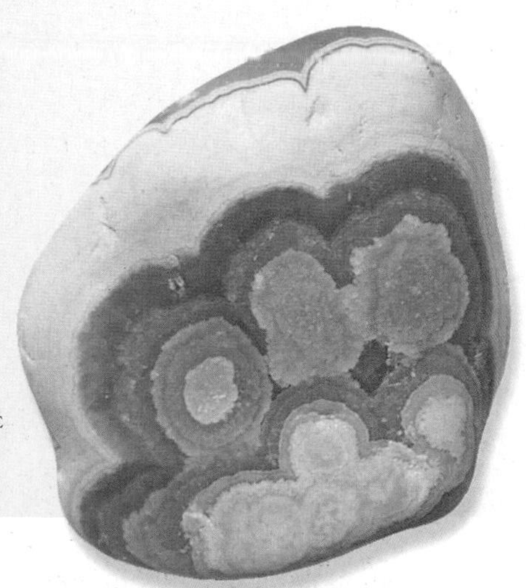

RHODOCHROSITE

GREEN JADE

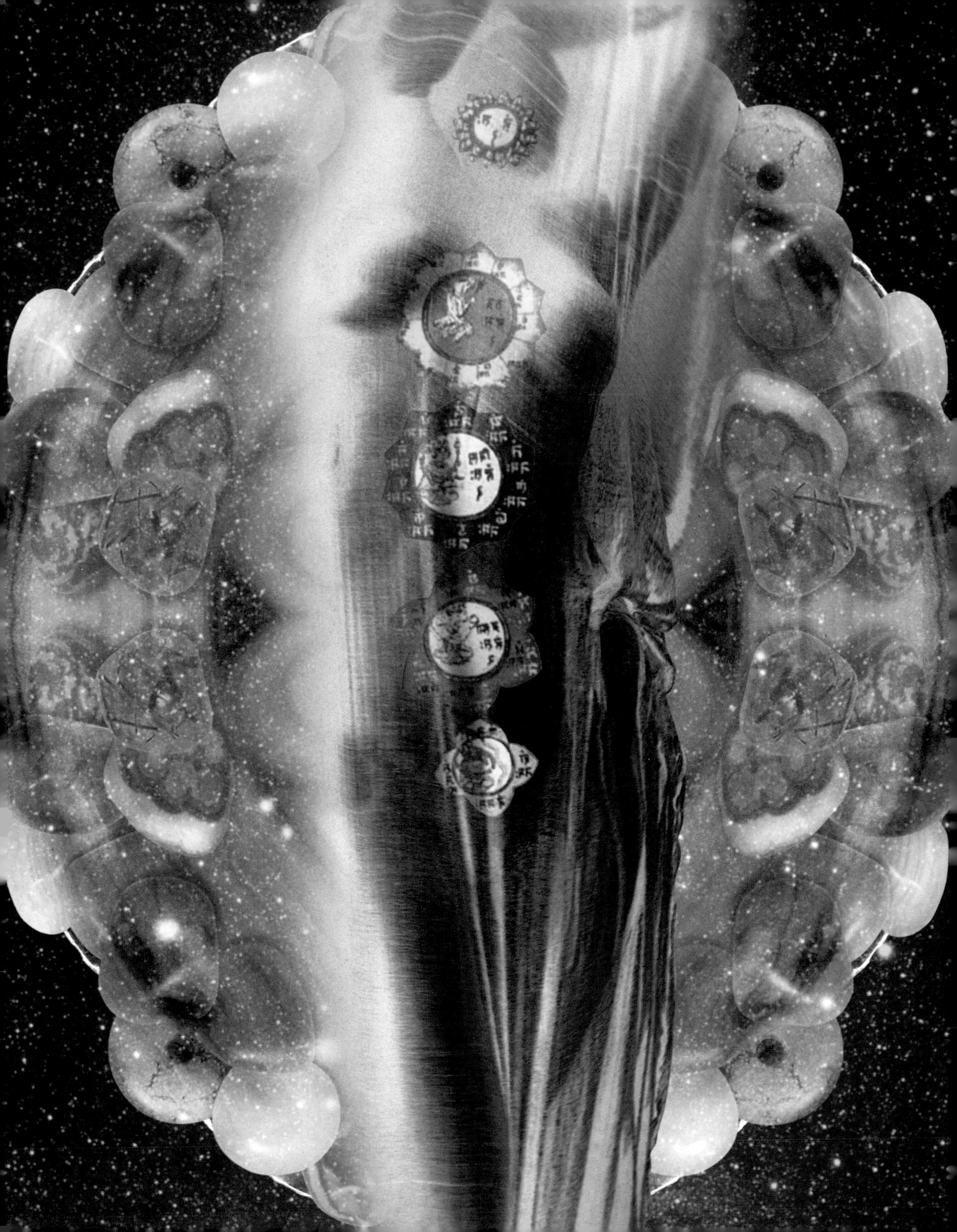

3 Crystal Healing

Crystals heal physically, emotionally, mentally, and spiritually, rebalancing and realigning energy to correct dis-ease. They act on the subtle, electromagnetic bodies surrounding the physical body and on the chakras – the linkage points connecting them.

When most people talk about "the body," they mean the physical body. However, present-day healers, like ancient shamans, recognize unseen bodies that emit subtle, electromagnetic energy and penetrate and surround the physical. They are referred to as "the aura" or "etheric body" (or bodies). A disturbance in one of these bodies, or a misaligned connection between it and the physical body, may result in dis-ease.

Many diseases are actually a state of dis-ease. The cause can be a combination of imbalances and energetic disturbances that, when they make themselves known on a physical level, cause the body to cease functioning properly.

Dis-ease can also occur on an emotional level. Too much energy manifests itself as anger, irritation, or frustration; too little energy as depression or apathy. If imbalances are mental, they could manifest as childhood conditions such as autism or dyslexia, or adult problems such as stress, mental confusion, or memory loss.

Spiritual dis-ease might manifest as a physical or emotional disturbance, or a disturbed mental state. Imbalances can be worked on through either the subtle bodies, the chakras, or the physical body. There are crystals to correct all and any energetic disturbances. They deal with causes not symptoms. Crystals are holistic because they work on all levels at once, creating a state of balance and wholeness.

Wands focus a crystal's innate healing energy. They can be rotated with a gentle circular motion over the appropriate part of the body.

Crystals and the chakras

Chakras are energy points linking the physical body with electromagnetic bodies around it. The name comes from the Sanskrit for wheel. If blocked or overactive, chakras create disharmony. If untreated, this leads to physical, emotional, or mental dis-ease. Crystals gently balance and realign chakra energies.

THE CHAKRAS

Chakras are energy centers that run up the spine to the crown of the head, linking subtle, electromagnetic bodies to the physical. Known as the aura or etheric body, subtle bodies surround the physical body, enfolding it within protective energy.

There are seven major chakras. There is also an earth chakra below the feet, a higher heart chakra just above the heart, and higher crown chakras above the head.

CHAKRA FUNCTIONS

The earth chakra keeps you grounded, holding you in everyday reality. If this chakra is not functioning there is difficulty with the material world.

The base and sacral chakras are the sexual and creative chakras. If the base chakra is blocked, libido is low, the creative spark nonexistent. If it is too active, sexual needs predominate. If the sacral chakra is blocked, social interaction can be difficult.

The solar plexus is where emotional blockages occur. If this chakra is too open, energy can be leached by other people.

If the heart chakra is blocked, love is prevented from flourishing. The higher heart chakra connects to unconditional, spiritual love. When this chakra opens, a new dimension in love is reached.

The throat is the chakra of communication. An open throat chakra allows free expression; a blocked one dams up communication of ideas and feelings.

The brow is where intuition arises and other realities can be accessed. If this chakra is too open, other people's thoughts and feelings flood in. If blocked, mental confusion arises and the imagination cannot function.

The crown chakra is where spiritual energies make themselves felt, and the higher crown chakras link up to angelic guidance. When this opens fully, there can be a feeling of bliss or enlightenment.

CHAKRA STONES

Chakras have crystals associated with them that cleanse, activate, or align their subtle energies. Specific color crystals can be used for chakra healing; these crystals can be selected to energize or sedate a chakra. Orange and red stimulate, blue and green calm, and violet elevates.

RIGHT Traditionally, there are seven major chakras:
- *The base chakra, at the bottom of the spine*
- *The sacral chakra just below the navel*
- *The solar plexus chakra, about a hand's width above the waist*
- *The heart chakra, over the physical heart*
- *The throat chakra, behind the voicebox*
- *The brow, or third eye, chakra, above and between the eyebrows*
- *The crown chakra, at the top of the head*

The ancient diagram shows these, although the top two are only barely visible.

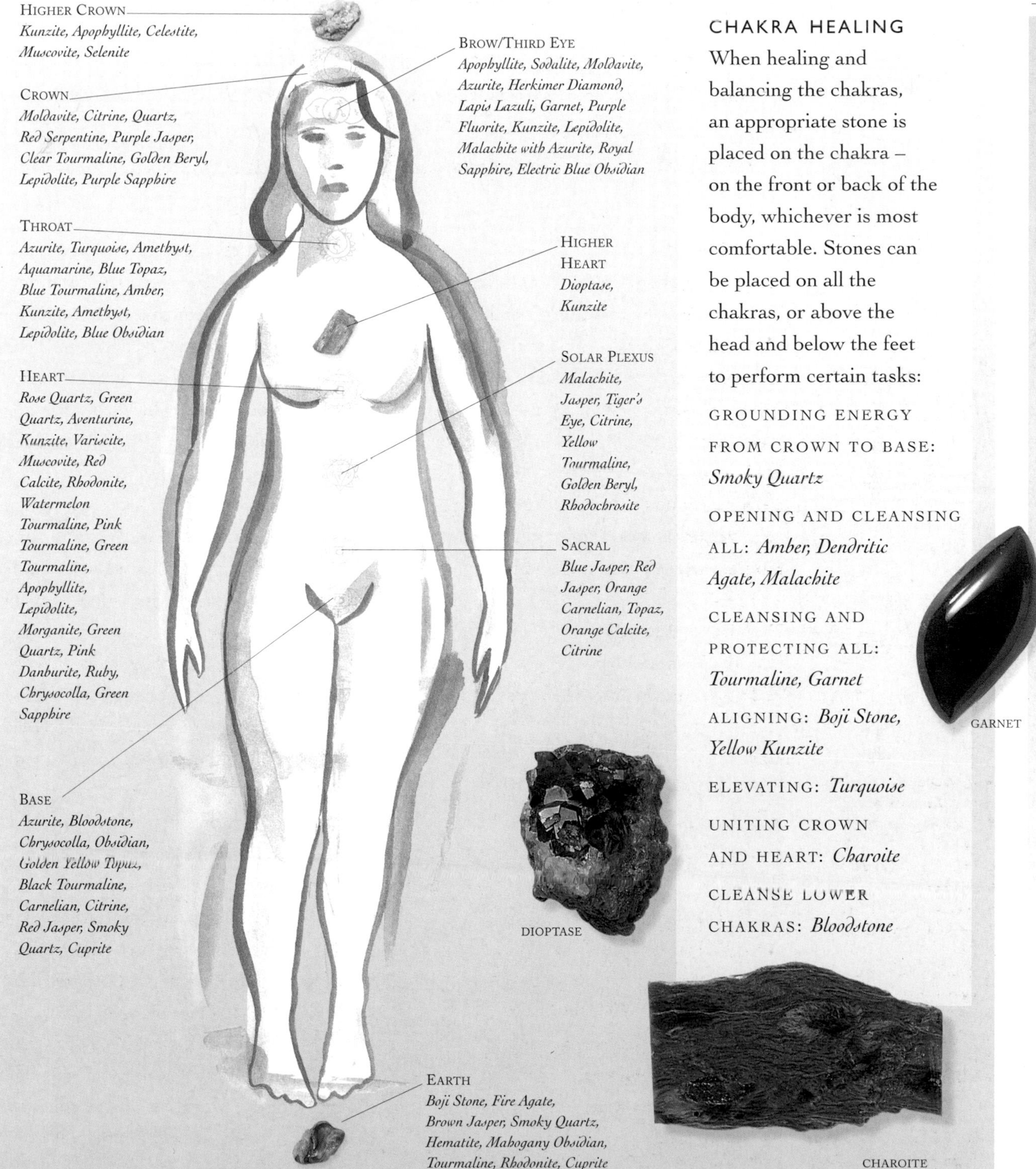

CHAKRA HEALING

When healing and balancing the chakras, an appropriate stone is placed on the chakra – on the front or back of the body, whichever is most comfortable. Stones can be placed on all the chakras, or above the head and below the feet to perform certain tasks:

GROUNDING ENERGY FROM CROWN TO BASE: *Smoky Quartz*

OPENING AND CLEANSING ALL: *Amber, Dendritic Agate, Malachite*

CLEANSING AND PROTECTING ALL: *Tourmaline, Garnet*

ALIGNING: *Boji Stone, Yellow Kunzite*

ELEVATING: *Turquoise*

UNITING CROWN AND HEART: *Charoite*

CLEANSE LOWER CHAKRAS: *Bloodstone*

GARNET

DIOPTASE

CHAROITE

IRON PYRITE

HEMATITE

ABOVE *The shape of a crystal affects its energy flow. Square crystals consolidate energy; rounded stones radiate energy.*

Crystals and the body

Crystals have an affinity with different parts of the body and can be used to correct imbalances in organs and bones. This resonance may arise from traditional astrological correspondences, or from the qualities and function of the stone itself. These connections are used to balance and heal the body. Appropriate stones can be placed on the body or made into elixirs for internal use.

CRYSTAL CONNECTIONS

The connections between crystals and the physical body have been known about for thousands of years. In Traditional Chinese Medicine (TCM), which goes back at least 5,000 years but is still in use today, a formula for ear problems contains Magnetite. In modern crystal healing, Magnetite has an affinity with bones but in the ancient Chinese system, both the bones and the ears are ruled by the kidneys – which Magnetite is said to stimulate. Modern crystal healing and TCM use Hematite for blood conditions and insomnia. Traditionally, it is said to calm the spirit, thus aiding sleep, and to cool the blood, arresting bleeding. The Chinese still use Iron Pyrite for bones and fractures, as do crystal healers today.

RIGHT *Traditional Chinese Medicine has always used crystals as part of its pharmacopeia. In China, physicians were paid only if their patient remained healthy, so healing methods had to be effective.*

Crystal formulas for medicines and unguents go back thousands of years. The ancient Egyptians left behind prescriptions that include Malachite, Chrysocolla, Jasper, Hematite, and Chalcedony. Malachite and Chrysocolla were ground up and sprinkled on wounds, and indeed modern science has proved that they are an effective bacteriocide, killing staphylococcus infections.

Greek and Roman healers such as Pliny and Galen continued this tradition. Galen used Hematite for headaches; Pliny used it for blood disorders. In medieval times, crystals formed an important part of a physician's pharmacopeia, and respected healers, such as the nun Hildegard of Bingen, utilized a wide range of gem elixirs in medicinal work. North and South American Native people used crystals for diagnosis and treatment.

Many crystal formulations have passed into modern medicines. However, crystal healers work by placing appropriate stones on the body, or preparing an elixir to drink or bathe the affected part. This means that substances that could be toxic if taken internally can be beneficially applied externally.

Crystal affinities

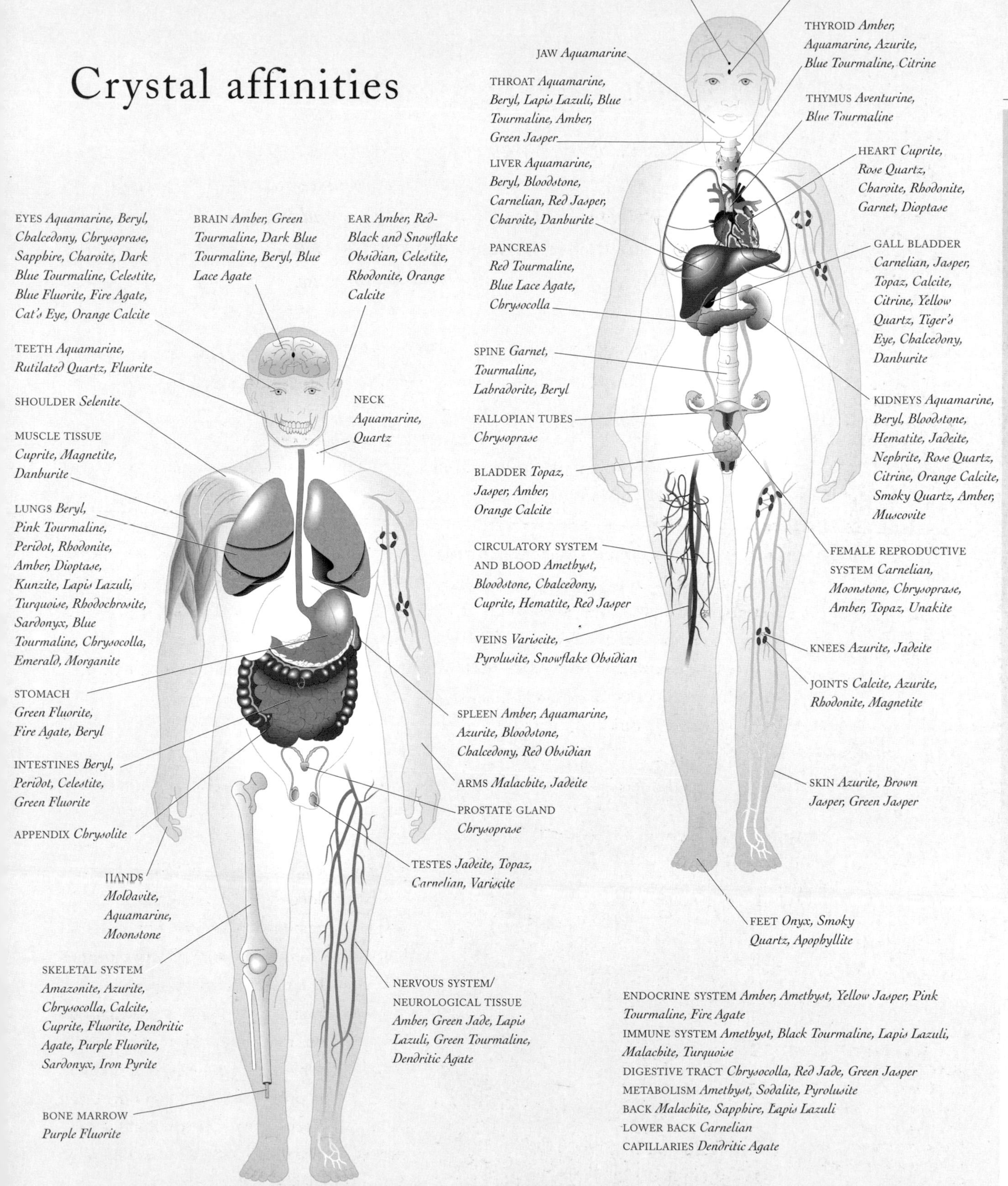

Healing physical conditions

Crystals are used for healing physical conditions in a variety of ways. Carrying a crystal that resonates with a physical condition can gently ameliorate the condition. Crystals can be placed, alone or in combination, on an affected body part or over the organ concerned. They can be swept over the body, which not only has an effect on a physical level but also rebalances subtle bodies. Alternatively, they can be placed in the bath.

RIGHT *Citrine radiates the healing and warming energy of the sun. Its effect is envigorating and detoxifying.*

Care must be taken when using crystals in combination, because some support each other and enhance healing, while others cancel each other out. Certain crystals bring conditions to a head quickly, activating a healing challenge, while others work more slowly. A stone like Magnetite, which has a positive and negative charge, will sedate an overactive organ, or stimulate a sluggish one.

BELOW *Lapis Lazuli has been used to heal painful headaches for at least 5,000 years. It is particularly effective for migraine.*

A crystal can be chosen on the basis of symptoms, especially if there is one single crystal that is associated with several of your symptoms. However, symptoms are merely that: symptomatic of a deeper dis-ease, an energetic imbalance in your body. A particular crystal could exacerbate a condition if its function is not fully understood. Consulting a qualified crystal healer is wise, because the healer will understand the energetic implications of your condition and how it resonates with appropriate crystals. The root cause of your dis-ease will be treated, rather than symptoms.

PAIN RELIEF

Pain calls your attention to the fact that something is wrong in your body, be it an excess of energy, a blockage, or an imbalance. Crystals gently sedate or release the energy, relieving the pain. Placed against the site of pain, a crystal draws it off. Cool and calming crystals, such as Lapis Lazuli, Turquoise, and Rose Quartz, sedate energy. More active crystals, such as Carnelian and Magnetite, remove blockages.

The metal copper has always been known for its ability to reduce inflammation and swellings. Malachite has a high concentration of copper and can relieve inflammation and draw off negative energies or imbalances. Aching muscles and joints benefit from Malachite and Magnetite. If muscles are knotted, dark Tourmaline can be beneficial. If the pain produces fear, a calming crystal such as Rose Quartz can be placed on the solar plexus.

HEADACHES

Headaches arise from a number of causes but are often related to stress or food. Amethyst, Amber, or Turquoise placed on or around the head will help relieve a tension headache. If the pain arises from food, a stone that calms the stomach may be needed such as Moonstone or Citrine. Lapis Lazuli has been used for migraine for centuries.

ABOVE *A crystal by your bed or slipped under your pillow will bring restful sleep. Select your crystal according to the cause of your insomnia.*

INSOMNIA

Another condition that may have several causes is insomnia. When sleeplessness is stress-induced, a crystal such as Chrysoprase, Rose Quartz, or Amethyst placed by the bed or under the pillow calms and soothes. If triggered by overeating, Iron Pyrite or Moonstone sedates the stomach. Where nightmares cause restless sleep, protective stones such as Tourmaline or Smoky Quartz promote peaceful sleep.

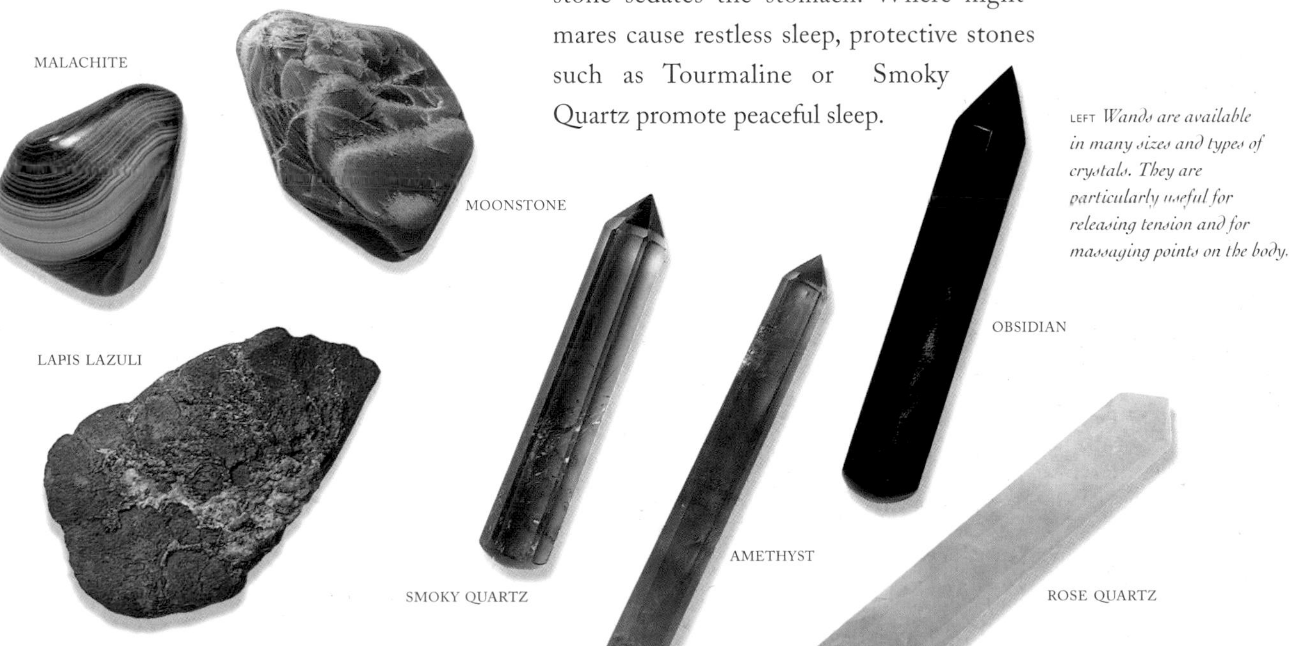

LEFT *Wands are available in many sizes and types of crystals. They are particularly useful for releasing tension and for massaging points on the body.*

Crystals to aid physical conditions

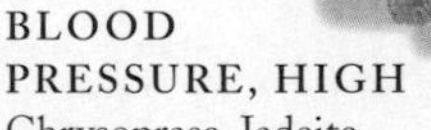

GREEN TOURMALINE

In case of serious illness, always consult your medical practitioner

A

ACNE
Amethyst (elixir)

ADDICTIONS
Amethyst, Kunzite

ALTITUDE SICKNESS
Cuprite

ALZHEIMER'S
Chalcedony, Blue Obsidian

ANEMIA
Bloodstone, Citrine, Kunzite, Tourmaline, Ruby, Tiger's Eye

ANGINA
Dioptase, Emerald

ANOREXIA
Rose Quartz

CITRINE

APPENDICITIS
Chrysolite

ARTHRITIS
Amethyst (elixir), Azurite, Blue Lace Agate, Black Tourmaline, Carnelian, Chrysocolla, Fluorite, Malachite, Rhodonite, Garnet

ASSIMILATE CALCIUM AND MAGNESIUM
Serpentine, Yellow Kunzite

ASSIMILATE IRON
Rhodonite

ASSIMILATE MINERALS
Chalcedony, Blue Jasper

ASSIMILATE PROTEIN
Opal

KUNZITE

ASSIMILATE VITAMINS AND MINERALS
Garnet

ASSIMILATE VITAMINS A AND E
Blue-Green Obsidian

ASTHMA
Amber, Amethyst, Malachite, Magnetite, Rose Quartz, Dark Blue Sapphire, Morganite, Azurite, Tiger's Eye

B

BACKACHE
Hematite, Magnetite, Malachite, Sapphire

BALANCE ORGAN FUNCTION
Magnetite

BILE DUCT
Jasper

BIRTH CONTRACTIONS, STRENGTHEN
Peridot

BLACKOUTS
Lapis Lazuli

BLADDER
Amber, Jasper, Orange Calcite

BLEEDING, STOP
Bloodstone, Ruby, Sapphire

BLOATING, DISPERSE
Green Jasper

BLOOD, CIRCULATION
Fire Agate

BLOOD, CLEANSING
Amethyst, Aquamarine, Bloodstone, Garnet, Lapis Lazuli

BLOOD CLOTS
Amethyst, Bloodstone, Hematite

BLOOD, OXYGENIZE
Amethyst, Carnelian

BLOOD POISONING
Carnelian

BLOOD PRESSURE, BALANCE
Aventurine, Charoite, Tourmaline

BLOOD PRESSURE, HIGH
Chrysoprase, Jadeite

BLOOD PRESSURE, LOW
Sodalite, Tourmaline

BONE MARROW
Violet Fluorite, Onyx

BONES, STRENGTHEN
Calcite, Onyx, Fluorite, Selenite, Sardonyx, Iron Pyrite, Amazonite

BOWELS
Jasper

BRAIN, IMPROVE FUNCTION
Lapis Lazuli

BRAIN FLUID, BALANCE
Blue Lace Agate

BREATHLESSNESS
Amber, Amethyst, Magnetite, Morganite

BRONCHITIS
Rutilated Quartz, Pyrolusite

BURNS
Quartz (clear), Rose Quartz (place in cold water)

BURSITIS
Amber, Blue Lace Agate

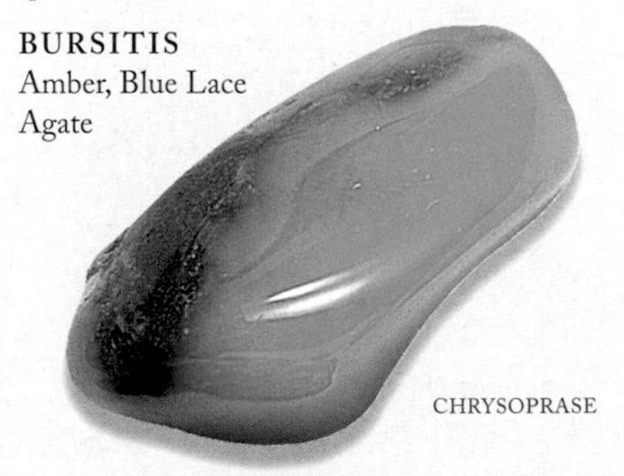

CHRYSOPRASE

C

CANDIDA ALBICANS
Carnelian

CATARRH
Topaz

CHEMOTHERAPY
Smoky Quartz, Herkimer Diamond

CHICKENPOX
Azurite, Malachite, Topaz

MOONSTONE

CHILDBIRTH, TO AID
Moonstone, Amber, Lapis Lazuli

CHROMOSOME DAMAGE
Chiastolite

COLDS/FEVERS
Jet, Emerald

COLIC
Carnelian

COMPUTER STRESS
Fluorite

CONCUSSION
Beryl

CONSTIPATION
Amber, Ruby

COUGH
Amber, Topaz

CRAMP
Bloodstone

TOPAZ

DEBILITATING ILLNESS
Black Tourmaline

DETOXIFICATION
Charoite, Jade, Peridot

DIABETES
Citrine, Jade, Serpentine

DIARRHEA
Quartz (clear), Malachite

DYSLEXIA
Sugilite

SUGILITE

E

EARACHE
Amazonite, Amber, Celestite, Tourmaline

ECZEMA
Sapphire

EMPHYSEMA
Amber, Amethyst, Dioptase, Malachite, Rhodonite, Rose Quartz, Tiger's Eye, Morganite

ENDOCRINE BALANCE
Amber, Amethyst, Tourmaline, Jasper, Citrine, Fire Agate, Green Quartz

ENVIRONMENTAL POLLUTION
Brown Jasper

EPILEPSY
Jasper, Lapis Lazuli, Sugilite, Tourmaline

EXHAUSTION
Yellow Jasper

EXTERNAL GROWTHS
Blue Lace Agate

EYES, INFLAMED
Blue Lace Agate (elixir), Sapphire, Chrysoprase

EYES, WATERING
Aquamarine

EYESIGHT, IMPROVE
Aquamarine, Charoite, Jade, Malachite, Rhodochrosite, Rose Quartz, Variscite

FALLOPIAN TUBES
Chrysoprase

FEET, BURNING
Blue Lace Agate, Onyx

FERTILITY, IMPROVE
Carnelian, Chrysoprase, Garnet, Jade, Malachite, Rose Quartz, Smoky Quartz

FEVERS
Red-Black Obsidian, Chiastolite, Ruby

FUNGAL INFECTIONS
Moss Agate (elixir)

BLUE LACE AGATE

GALL BLADDER
Carnelian, Citrine, Jasper, Topaz

GINGIVITIS
Blue Lace Agate (elixir as mouthwash)

GLANDS, SWOLLEN
Aquamarine, Blue Lace Agate, Topaz

GLANDULAR FEVER
Blue Lace Agate

GOITER
Amber

GOUT
Chrysoprase, Topaz, Tourmaline

GUMS
Pyrolusite, Agate

HAYFEVER
Blue Lace Agate

HEADACHE
Amber, Amethyst, Emerald, Lapis Lazuli, Turquoise, Sugilite (with Black Manganese), Charoite, Cat's Eye, Hematite, Citrine, Moonstone

HEARING LOSS
Rhodonite, Tourmaline

HEART ATTACK
Dioptase

HEARTBURN
Dioptase, Quartz (clear), Peridot

HEAT STROKE
Blue Lace Agate

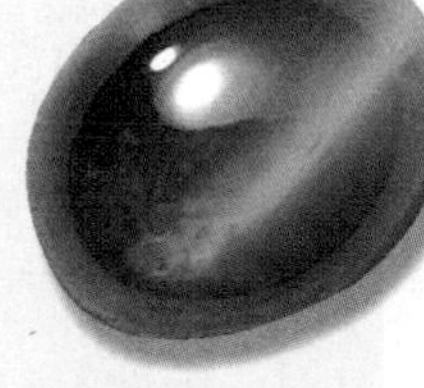
CAT'S EYE

Crystals to aid physical conditions

RHODONITE

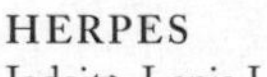

HERPES
Jadeite, Lapis Lazuli

HIP PAIN
Azurite

HIV AND AIDS
Jadeite, Lapis Lazuli, Amethyst

HORMONE PRODUCTION
Amethyst

HYDROCEPHALUS
Blue Lace Agate (elixir)

HYPERTENSION
Chrysocolla

CARNELIAN

I

IMMUNE SYSTEM, STRENGTHEN
Amethyst, Lapis Lazuli, Malachite, Jade, Quartz (clear), Tourmaline

IMPOTENCE
Amazonite, Variscite

INFERTILITY
Rose Quartz, Carnelian, Chrysoprase, Smoky Quartz, Jade

INSOMNIA
Amethyst, Lepidolite, Sapphire, Sodalite, Topaz, Chrysoprase, Lapis Lazuli, Citrine, Tourmaline, Iron Pyrite, Labradorite

INSULIN REGULATION
Chrysocolla, Opal

INTERNAL INFECTIONS (EAR, SINUS, ETC.)
Rhodochrosite (elixir, poultice), Opal

ITCHING
Azurite, Malachite

AZURITE

J

JAUNDICE
Jadeite

JOINT PROBLEMS
Azurite

K

KIDNEY DISEASE
Jadeite, Nephrite

KNEE PROBLEMS
Azurite, Green Jadeite

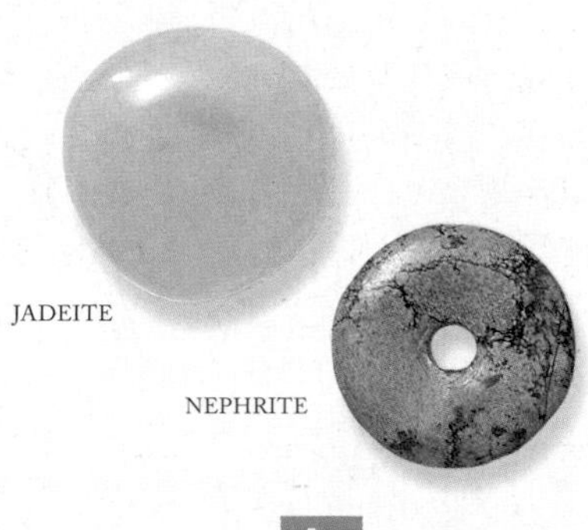
JADEITE

NEPHRITE

L

LACTATION
Chalcedony, Chiastolite

LARYNGITIS
Amber, Blue Lace Agate, Tourmaline, Sodalite

LEG CRAMPS
Hematite

LEUKEMIA, SIDE-EFFECTS
Chrysocolla

LIVER FUNCTION
Amethyst, Aquamarine, Beryl, Bloodstone, Charoite, Jasper, Jade, Topaz

LOWER BACK PROBLEMS
Carnelian

LUMBAGO
Magnetite

LUNGS, FUNCTION
Rhodochrosite, Chrysocolla

LUNGS, POLLUTION
Turquoise

LYMPHATIC FUNCTION
Tourmaline

M

ME
Ruby, Tourmaline

MEASLES
Turquoise

MÉNIÈRE'S DISEASE
Dioptase

MENOPAUSAL SYMPTOMS
Lapis Lazuli, Garnet, Ruby, Lepidolite

UNAK

MENSTRUAL PAIN
Rose Quartz, Lapis Lazuli

MENSTRUAL PROBLEMS
Jet, Moonstone, Unakite

METABOLIC IMBALANCES
Amethyst, Cuprite, Chrysocolla, Sodalite, Moonstone, Labradorite, Pyrolusite

MIGRAINE
Lapis Lazuli

MUMPS
Aquamarine, Topaz

MUSCLE CRAMPS
Chrysocolla, Dioptase, Hematite

MUSCLE STRAIN
Magnetite

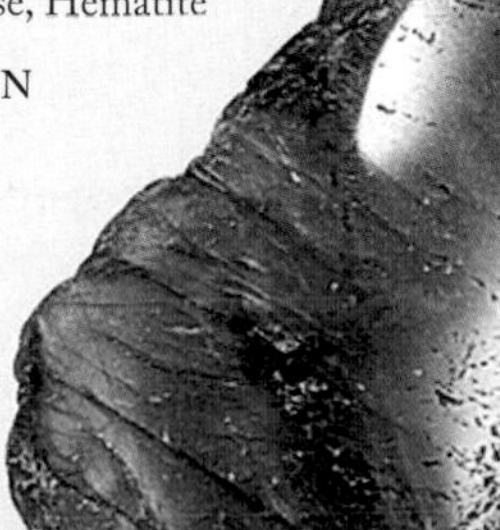
LABRADORITE

HERKIMER DIAMOND

NAUSEA
Jasper, Emerald

NEPHRITIS
Nephrite, Jadeite

NEURALGIA
Lapis Lazuli, Carnelian, Amber, Amethyst

NOSE BLEED
Carnelian

OBESITY
Green Tourmaline

OSTEOPOROSIS
Amazonite

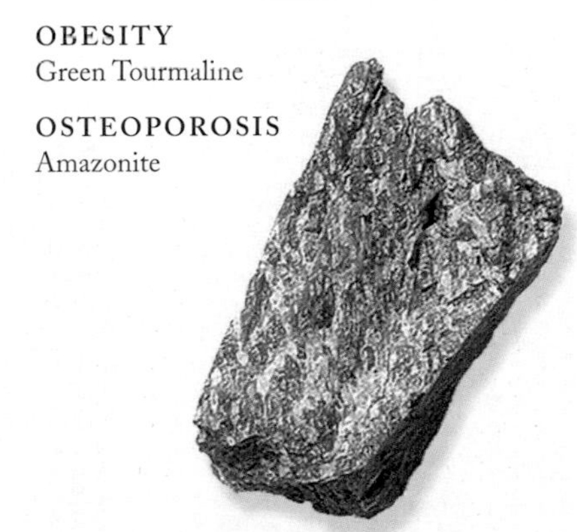
AMAZONITE

PAIN RELIEF
Lapis Lazuli, Magnetite, Dioptase, Rose Quartz, Turquoise, Carnelian, Malachite, Carnelian

PARKINSON'S DISEASE
Opal

PITUITARY GLAND IMBALANCE
Benitoite

PNEUMONIA
Fluorite

PROSTATE GLAND
Chrysoprase

PSORIASIS
Blue Lace Agate, Labradorite

R

RADIATION-RELATED ILLNESS/THERAPY
Smoky Quartz, Herkimer Diamond, Yellow Kunzite, Malachite, Sodalite

RHEUMATISM
Agate, Amber, Azurite, Carnelian, Chrysocolla, Fluorite, Malachite

SCAR TISSUE
Rose Quartz

SCIATICA
Tourmaline, Sapphire

SEXUALLY TRANSMITTED DISEASES
Chrysoprase

SEXUAL LIBIDO, TO REKINDLE
Fluorite

SINUS
Azurite, Blue Lace Agate

SKIN PROBLEMS
Aquamarine (elixir)

SORES
Chalcedony

SORE THROAT
Amber, Aquamarine, Beryl, Lapis Lazuli, Blue Tourmaline

SPINAL ALIGNMENT
Azurite, Hematite, Labradorite, Magnetite, Tourmaline, White Jade

SPORTS INJURIES
Magnetite

THYMUS, STIMULATE
Aventurine, Dioptase, Quartz (clear)

THYROID, BALANCE
Rhodochrosite (elixir), Citrine

THYROID, SEDATE
Turquoise, Sodalite

TISSUE REGENERATION
Peridot

TOOTHACHE
Amber, Aquamarine, Lapis Lazuli

TOOTH DECAY
Amazonite, Fluorite, Onyx

TRAVEL-SICKNESS
Jasper

TUBERCULOSIS
Morganite

TUMORS
Amethyst, Bloodstone, Smoky Quartz

LAPIS LAZULI

ULCERS
Chrysocolla

VARICOSE VEINS
Blue Lace Agate, Bloodstone, Amber

VERTIGO
Quartz (clear), Cuprite

VIRILITY, IMPROVE
Lapis Lazuli, Jet, Black Quartz, Red-Black Obsidian

VOMITING
Lapis Lazuli

WATER RETENTION
Cuprite

WHOOPING COUGH
Topaz, Amber, Blue Lace Agate

AQUAMARINE

Healing emotional imbalance

Emotions that are out of balance create subtle dis-ease, while repressed emotions can lead to depression, phobias, and fears. Crystals facilitate achieving emotional equilibrium. They dissolve blockages to emotional expression and help promote peace and tranquillity.

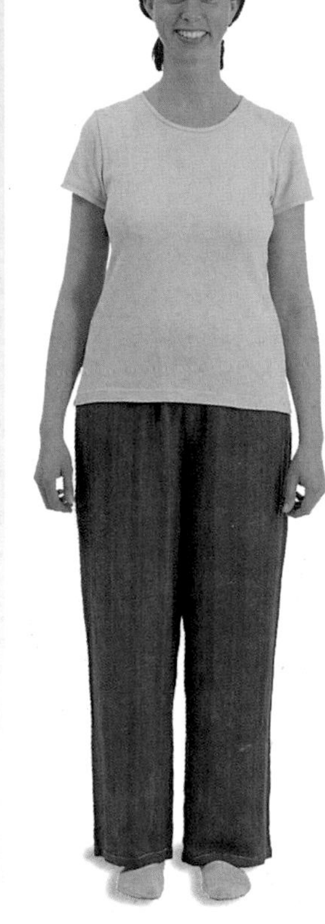

ABOVE *Balanced emotions contribute significantly to your health and well-being.*

BALANCING EMOTIONAL STATES

Blue Lace Agate and Rose Quartz are great for cleansing and detoxifying the emotions. Opal promotes emotional balance and stability, and Amethyst, too, balances emotional highs and lows, working on hormone production and helping you to feel less scattered and more in control. The stones should be worn or kept about your person. They can also be placed in a healing layout or used in the bath.

DEPRESSION

Amber neutralizes negative states of mind and balances the underlying emotional and endocrine disequilibrium. Its sunny yellow color dispels gloom and instills positive energy. Jet draws out and absorbs negative energy, controlling mood swings and fighting deep depression.

Deep Blue Sapphire brings serenity, helping to reach your true nature. Knowing and accepting yourself neutralizes depression. Chrysolite and Smoky Quartz also relieve depression and negative states of mind, helping you feel less burdened. Post-natal depression is lifted by Rose Quartz. Where depression is one pole of mood swings, then Kunzite helps to stabilize the condition.

ANGER

Anger can be debilitating if it is blocked or gets out of control. Suppressed anger often underlies depression and addiction. Agate clears bitter anger that eats away deep within oneself, and Peridot, too, aids this process. Red Garnet helps reduce unreasonable anger, especially toward yourself, while Red Jade aids constructive expression of anger. Moonstone and Amethyst soothe emotional overreactions but are particularly useful for ameliorating anger, especially toward the self. If addictions are a problem, Amethyst and Kunzite can overcome them.

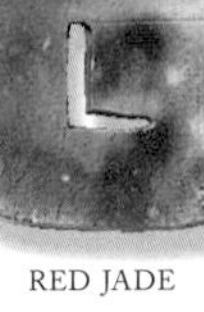

RED JADE

JET

AMBER

KUNZITE

GUILT

Guilt is a disabling emotion. Chrysocolla clears guilt or fear, making you more comfortable with speaking the truth. The loving energy of Rose Quartz helps to dissolve guilt, allowing you to love yourself once more.

LACK OF LIBIDO

Lack of libido frequently has an emotional cause. Fluorite and Red Garnet can gently release emotional conditions blocking sexual energy, and stimulate passion once more.

PHOBIAS AND FEARS

Aquamarine has the ability to stabilize and harmonize emotional energy. It dispels fear and banishes phobias. If anxiety is the problem, Aventurine, Green Calcite, Chrysoprase, Kunzite, Iron Pyrite, and Tourmaline alleviate it. The glassy volcanic rock Obsidian is a double-edged sword. It brings things to the surface rapidly and can release inner fears, but the shadow qualities revealed are often believed to be negative and undesirable. This is a stone to use when you are ready to change something about yourself. It may need to be followed by Rhodochrosite or Rose Quartz, bringing in positive qualities to replace the negative, encouraging self-love.

SELENITE

FORGIVENESS

One of the most effective ways of healing emotional imbalance is to practice forgiveness, especially self-forgiveness. Chrysoberyl, Rhodochrosite, Rose Quartz, and Selenite aid this process, while Celestite restores emotional innocence.

IRON PYRITE

ROSE QUARTZ

CHRYSOBERYL

CHRYSOPRASE

BLUE FLUORITE

GREEN CALCITE

Healing mental states

There are close links between the nervous system and the mind. The brain and the nervous system process information and act on it. If the nervous system is out of balance, the mind will be disturbed. There are also links to emotional states. When depression is present and self-esteem low, confidence will be lacking and confusion prevails. Balancing these states and encouraging positive qualities brings harmony to the mind.

AMETHYST

Crystals for mental healing can be worn as earrings or pendants or placed around the head while in a relaxed state. Holding one of the mentally calming stones is helpful in times of mental pressure. An appropriate crystal on the desk or in the pocket can aid concentration and mental focus, calming the nerves. Adding a soothing crystal to bath water also aids mental stress.

BELOW *Earrings are a perfect way to wear stones that keep your mind free from stress.*

CRYSTALS FOR HEALING THE MIND

Green is a soothing color, and many green crystals reduce mental and nervous stress. Green Jade is much prized in the East for its ability to calm the nervous system and to focus the mind.

Amazonite filters the information that reaches the brain, preventing mental overload. It soothes the nervous system and helps you to find the most effective way of expressing yourself, releasing nervous tension. Aquamarine also filters mental information, sharpening perception and aiding clear communication. It brings about a serene state of mind and creates clarity.

Purple relaxes the mind, raising it to a different plane. Amethyst is a powerful mental healer, making you less scattered and more in control of your mind. It soothes the nervous system, aids neural transmission, focuses realistic goals, and boosts memory. It helps you to visualize clearly. If mental stress is creating insomnia, Amethyst gently lulls you to sleep. You can use Amethyst in a healing layout around your body. Place one crystal above your head and one below your feet. Place six other crystals on each side of the body at shoulders, waist, and knees. If the crystals have points, face them in toward the body. Spend 20 minutes lying quietly. This layout is so relaxing that when you have finished, you may need a grounding stone such as Smoky Quartz to bring you back to earth.

OVERCOMING MENTAL STRESS

By soothing the mind and filtering information, Amazonite helps focus attention on what is important. Beryl clears mental stress by teaching you how to cease doing the unnecessary. It filters out distractions and excessive stimulation, but stimulates the intellect. Amethyst, too, relieves stress by

reducing mental burdens and helping you to focus on realistic goals. With Amethyst-enhanced neural transmission and mental clarity, the body is able to act on information that comes from the mind.

If mental stability and balance are needed, Iron Pyrite, Aquamarine, Beryl, and Green Calcite support the mind, while Rhodochrosite prevents mental breakdown. Bloodstone, Dioptase, Hematite, Lapis Lazuli, Rhodonite also soothe mental stress.

FOCUSING THOUGHTS

Many people experience difficulty in focusing and controlling their thoughts. The mind flits from one topic to another, multitudes of thoughts interfere with concentration, and information gets lost in the mental maze. Enhancing study and concentration, Quartz gives mental clarity and focus, and Carnelian clears extraneous thoughts.

Azurite helps a person to attain a meditative state and to gain control of the mind. It is said to have the capacity to rebuild brain cells. This copper-based stone was used by the priests and priestesses of ancient Egypt to enhance their spiritual consciousness. Azurite promotes objectivity, insight, and clear thought.

If poor memory is getting in the way, Citrine, Amber, or Yellow Fluorite stimulate memory, while Green Calcite sharpens mental clarity and boosts memory. Blue Calcite calms thoughts, allowing them to become focused. Lapis Lazuli is a powerful amplifier of thoughts. Sapphire is a useful stone for mental activity, alleviating confusion. Apophyllite brings balance when there is excessive energy in the head.

BELOW *This ancient Egyptian statue is covered with magical texts for invoking healing. Such invocations would support the practical work of the physician, who used a wide variety of highly effective crystals as medicine.*

ABOVE *Placing eight Amethyst points around your body brings relaxation and mental equilibrium.*

FLUORITE

Citrine, Hematite, Lapis Lazuli, Red Garnet, Rhodonite, Rose Quartz, Ruby, Tourmaline, and Variscite all aid concentration, and Emerald, Fluorite, Moss Agate, Rhodochrosite, and Tourmaline banish forgetfulness.

EMERALD

RHODONITE

PROBLEM-SOLVING

Problem-solving needs both creative analysis and intuition. Lateral thinking – looking at problems from an unexpected angle – often reveals a solution, and for this Green Tourmaline is particularly useful. It helps you to analyze difficulties and follow a problem through to its conclusion, stimulating communication along the way. Aquamarine's ability to filter out unnecessary information also enhances problem-solving, particularly because it aids communication of insights. Peridot, a mental cleanser, highlights problems, and Obsidian, if used circumspectly, brings up the hidden factors that block problem-solving. If assistance is needed in making a decision, Azurite and Rutilated Quartz will come to your aid.

IMPROVING COMMUNICATION

If mental processes are woolly, then communication is unclear. Some people have brilliant ideas but cannot organize or express them, so Sodalite and Sapphire would help enhance communication skills. Sodalite makes you less critical of yourself and others, creating objectivity, and helps you to change your attitude. Aquamarine unblocks communication, aiding verbal expression, filtering out unnecessary information, and increasing perception. It is a useful stone for those involved in public speaking.

SODALITE

SAPPHIRE

AQUAMARINE

Intuition boosters include Azurite, Chrysocolla, Emerald, Smoky Quartz, Sodalite. Creativity is enhanced by Agate, Amazonite, Aquamarine, Aventurine, Azurite, Carnelian, Celestite, Chalcedony, Chrysocolla, Citrine, Quartz, Herkimer Diamond, Iron Pyrite, Sardonyx, and Topaz. Turquoise, Chiastolite, and Muscovite facilitate problem-solving.

LEARNING DIFFICULTIES

Sugilite is a most useful stone for anyone with autism or learning difficulties such as dyslexia because it aids mental coordination. Sugilite can help with accepting difference. It highlights the lessons being learned while in a physical body, softening them with love.

If intellectual stimulation is required, then Yellow Fluorite, Gold Calcite, and Sapphire can all help.

LEFT *Crystals can have a powerful effect on mental disturbances of all kinds. Some of the crystals traditionally used in treating depression, such as those containing lithium, form the basis for present-day psychiatric drugs.*

PSYCHIATRIC DISORDERS

Chrysoprase has been used for hundreds of years in the treatment of mental illness. It gently calms and opens the mind to new experiences. Kunzite contains lithium, which is used in psychiatry to control mental disorders such as bipolar affective disorder. It can help in making adjustment to the pressures of life, and shields the person from undue mental influence from outside.

If delusions are a problem, Carnelian gently dispels them, and Lapis Lazuli will do the same for hallucinations.

BELOW LEFT *Sugilite helps both those who have learning problems and their teachers, bringing love and understanding of difficulties.*

CHALCEDONY

SUGILITE

GOLD CALCITE

Gem remedies

Gem remedies, or elixirs, are made by placing a crystal in pure spring water and exposing it to sunlight. The crystal transfers its subtle vibrations to the water. The water is taken internally or used to bathe the affected part.

ABOVE *Dark glass bottles best protect the energies of elixirs, but any clean, airtight bottle can be used.*

Elixirs work on physical, emotional, mental, and spiritual states. In ancient Egypt, crystals would be left out overnight and the dew carefully collected. Crystal bottles would be filled with water, left in sunlight, and the water taken internally or used as a bath.

ABOVE *Hildegard of Bingen was famous for her healing work. She left behind many recipes for gem elixirs.*

In the Middle Ages, the healer-nun Hildegard of Bingen prepared many elixirs, including one of Topaz for eyes. The stone was placed in wine for three days, then the Topaz was touched to the eyes at night. After five days, the wine could be drunk. Wine was used rather than water because of the risk of contamination from the poor-quality drinking water.

One of the primary uses of gem elixirs is to balance and stabilize the chakras and the aura that surrounds the physical body. They release stress and dissonant energies, allowing the subtle bodies, the chakras, and the physical body to come into alignment. They also help the physical body to integrate healing changes.

CROWN CHAKRA

BASE CHAKRA

SACRAL CHAKRA

HEART CHAKRA

MAKING AN ELIXIR

Nowadays, ready-made gem remedies are available for a wide variety of purposes, but you can also make your own. Gem remedies are prepared by placing a crystal into pure spring water, in a glass bowl or pitcher, and placing it in sunlight to energize and activate. The crystal passes its subtle healing vibrations to the water. It is then bottled in a clean, glass bottle, and drunk or used to bathe the affected part. Brandy can be added as a preservative.

Crystals for elixirs should be cleansed before use. Care needs to be taken because some crystals could be toxic. Malachite, for instance, should only be used in its polished form. If in doubt, apply externally only.

LEFT *Store your remedies in a cool dark place. Brandy, which acts as a preservative, can be added if the remedy is not to be used immediately.*

GOLDEN BERYL

USEFUL GEM REMEDIES

Gem	Remedy
BLACK TOURMALINE	*Provides psychic protection and screens from electromagnetic stress. Relieves jet lag. Releases toxic energy from emotions, mind, and body.*
MALACHITE	*Harmonizes physical, mental, emotional, and spiritual; grounds the body.*
FLUORITE	*Breaks up blockages in the etheric body.*
JADEITE	*Heals eye conditions, brings peace.*
AMAZONITE	*Balances the metabolism.*
GREEN JASPER	*Restores biorhythms and natural sexuality.*
HEMATITE	*Strengthens boundaries.*
KUNZITE	*Opens the heart.*
AMBER	*Acts as an antibiotic, heals throat problems.*
GOLDEN BERYL	*A gargle for sore throats.*
BLOODSTONE	*Releases constipation and emotional stagnation.*
CHAROITE	*An excellent cleanser for the body.*
HERKIMER DIAMOND	*Aids psychic vision and dream recall.*
MOSS AGATE	*Treats fungal infections.*
MOLDAVITE	*Aids staying in the present moment, releasing from past.*
AQUAMARINE AND AVENTURINE	*Aids in the case of skin problems.*
AQUAMARINE	*Calms mental stress and overactive mind.*
AVENTURINE	*Increases stamina to see things through, integrates new experiences.*
RHODOCHROSITE	*Treats infections, promotes thyroid balance, heals the heart.*
TIGER'S EYE	*Aids self-empowerment.*
SAPPHIRE	*Removes toxins from the body.*
BLUE LACE AGATE	*Balances brain fluid, mouthwash for gingivitis, bath for inflamed eyes.*
AMAZONITE	*Balances metabolism.*
AMETHYST	*Treats acne and arthritis.*
QUARTZ	*An energy enhancer and general cleanser.*
CARNELIAN	*Reenergizes.*
CITRINE	*Clears mental confusion, improves concentration and decision-making.*
CHRYSOCOLLA	*Releases unresolved grief.*

If in doubt, consult a qualified crystal therapist.

BLACK TOURMALINE

TIGER'S EYE

MOSS AGATE

CHRYSOCOLLA

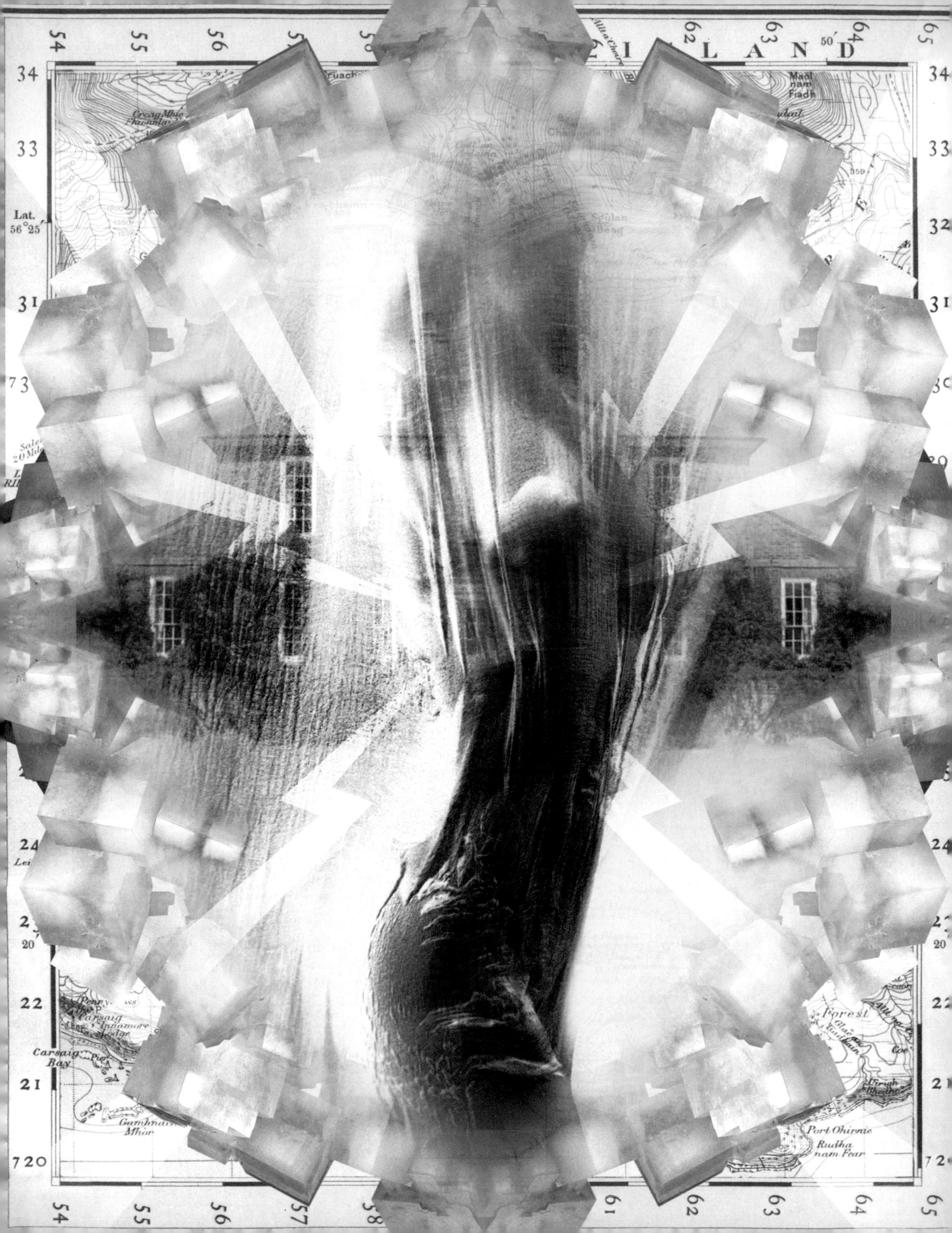
LAND
Maol
nam
Fiadh
Lat.
56°25'
Carsaig
Bay
Forest
Port Ohirnie
Rudha
nam Fear

4 Crystal Protection

No matter how rural the environment, the air is full of unseen energies. Cables carry electricity, antennas beam radio and television signals and microwaves. In a city, subtle-energy environmental pollution is even stronger. The earth itself carries negative forces, such as water lines, which create geopathic stress.

Houses have a subtle-energy history. An imprint is left by people who live there. Neighbors add their vibes. At home and work, computers and other equipment generate electromagnetic fields that create fatigue and irritability. Strong electromagnetic fields are implicated in illnesses such as cancer and leukemia. Energies are transmitted at frequencies invisible to the naked eye. Special cameras pick up the infrared spectrum, meters measure electromagnetic radiation and microwave output, but there is no way to measure the effects of negative thoughts and toxic emotions. Other people's negativity can deplete your energy if you pick up their thoughts and feelings. Family members can create stress in each other. Sometimes you can be unlucky enough to provoke deliberate ill-wishing (known as psychic attack).

Wearing crystals means that their protective power is close to you at all times.

Fortunately crystals offer a safe way to counteract environmental and personal pollution, and to protect yourself and your surroundings. A Fluorite or Quartz cluster by the computer neutralizes electromagnetics. Smoky Quartz absorbs geopathic stress, Turquoise clears environmental pollution, Black Tourmaline deflects psychic attack. Amber or Bloodstone cleanse negative energies, which can be replaced by the positive, loving vibrations of Rose Quartz or Amethyst. Place crystals near the party wall if you have noisy neighbors: it will calm them down!

ABOVE *Rose Quartz placed by a party wall calms noisy neighbors and brings peace and harmony to both houses.*

Protecting your space

Crystals are beneficial for your self, home, family, work, and external surroundings. Beautiful to look at, their energy gently deflects negativity and introduces positive energies that make you feel good. They absorb excess energy and boost depleted energy, bringing harmony and protecting your space.

ABOVE *A Lepidolite (or Purple Fluorite) crystal placed by your computer will absorb electromagnetic emanations.*

A crystal has benefits that you feel rather than see. Place it in the right spot and it brings protection and well-being. A protective stone outside the front door will guard the whole house, even from burglars. If you want your home filled with love and harmony, Rose Quartz does the trick.

How do you find the right spot? The crystal will probably tell you! Some places are obvious. If you are creating an electromagnetic-free zone, placing a Lepidolite, Quartz, or Fluorite cluster by the computer or television makes sense. A large piece of Rose Quartz by your bed encourages love and harmony. A crystal near the wall you share with noisy neighbors radiates out to them. Remember to clean the crystals regularly.

PROTECTING YOUR CAR

A crystal placed securely inside your car keeps you safe. Choose one of the protective crystals and program it to look after you and the car.

CREATING A SAFE WORK PLACE

Place a Smoky Quartz on your desk to protect you from other people's stress and frustration. Program it to create a protective bubble around you. If your job is stressful, you will find that holding a calming stone like Rose Quartz or Amethyst is helpful.

DOWSING

To dowse for the right spot you will need a crystal on a chain (a pendulum). Clear Quartz is ideal. Hold the chain between your thumb and fingers with about a hand's width of chain hanging down. It should feel comfortable.

To check for "yes" and "no," hold the pendulum over your hand. State your name and ask if that is correct. The crystal will swing in a particular way. This is "yes." Then state another name and ask if this is your name. The crystal will swing in a different way to indicate "no."

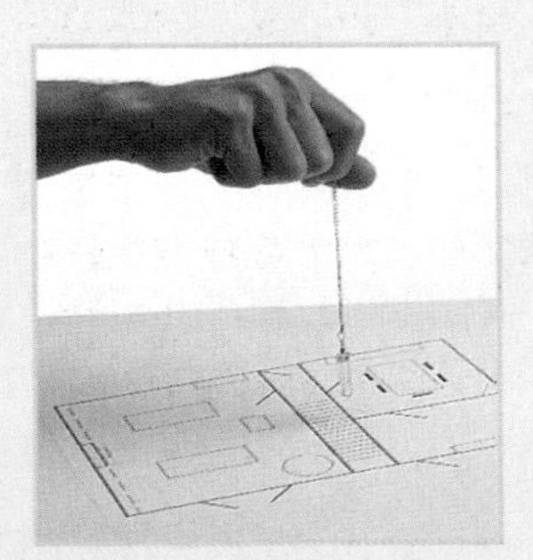

To place a crystal in your home, draw a sketch map of the room, ask where the right place is, and move the pendulum until you get a "yes." Alternatively, place the crystal in different parts of the room and check its position with the pendulum.

SAFE SPACE CRYSTALS

BLACK TOURMALINE, JET, SMOKY QUARTZ, BLACK JADE, AND FLUORITE
Absorb and deflect energies

TURQUOISE
Protects against pollution

AMETHYST AND ROSE QUARTZ
Attract loving energy

AMBER AND BLOODSTONE
Cleanse and transform negativity

LEFT *Some forms of Smoky Quartz – the dark one shown here on the far left – have been irradiated to darken their color. A natural stone like the one on the near left is more appropriate in dealing with environmental stress; an irradiated stone could exacerbate it.*

EXTERNAL ENVIRONMENT

The environment around your home may have "bad vibes." You can easily correct this with an appropriate crystal such as Turquoise or Tourmaline. If the crystal has a point, point it away from your own house to deflect negative energy or to transmute it into positive energy.

RIGHT *Place a reflective crystal outside your front door – with the point facing away – to deflect any negative energies rushing at your door.*

Nuclear radiation

If you live near a nuclear power station, or a hospital that practices nuclear medicine, crystals will protect you. Certain areas retain the effects of nuclear disasters, and some land has natural radiation. Malachite, Smoky Quartz, Herkimer Diamond, Yellow Kunzite, and Sodalite help absorb radiation and protect against high-frequency communication antennas, infrared, microwaves, and radar.

Protecting yourself and the aura

The subtle bodies that surround the physical, known as the aura, act as a protective screen. When they are functioning correctly, they repel other people's energy and protect against energy leakage, but they can be damaged by stress and other factors. Crystals heal the aura and provide protection.

APACHE TEAR

BELOW *Crystals can help to keep your aura functioning well. With a healthy aura, you are not susceptible to psychic invasion or energy leakage.*

AURIC PROTECTION

The aura is a field of colored light that surrounds the physical body. It is vital and energetic, active and vibrating – and invisible to the naked eye. It has different layers of multi-colored bands. The aura delineates "your space." If someone comes in too close and enters your auric field, you feel invaded. But someone need not come physically close to invade your space. If your aura is not strong, external energies, thoughts, and feelings can penetrate the aura, even from a distance. This is where auric protection comes in.

By strengthening the aura, you make it impossible for other people to penetrate your personal space. You no longer pick up their thoughts and feelings – their dis-ease. Nor can they leach your energy. Labradorite and Apache Tears are particularly useful as aura protectors. Fluorite, which provides a protective shield, is also good.

LABRADORITE

PSYCHIC VAMPIRES

Some people behave like psychic vampires. Instead of blood, they feed on energy. It is as though they pull out a plug and all your energy escapes. Everyone has someone whose company they enjoy but, when that friend has left, there is a feeling of exhaustion. The person is a psychic vampire! Fortunately you do not need to drop that person from your life. Take the right crystal precautions, and your energy stays with you. Green Aventurine shields you from vampirism of your heart energy, and Labradorite prevents energy loss. To take another approach, try Black Tourmaline – it will not allow a psychic vampire to get close enough to suck the energy from you.

There are several ways of working with crystals to strengthen the aura.

You can work through the chakras, the linkage points between the aura and the physical body. If your chakras are balanced and in good order, they will strengthen the aura and help to prevent energy leakage.

ABOVE *Amber is an excellent protector for the aura and makes an attractive piece of jewelry.*

Wearing an appropriate crystal is helpful, particularly when you are with other people. If you are the sort of person who gets on a bus feeling wonderful and gets off feeling dreadful, then you need auric protection. Wear the right crystal and you need never suffer psychic depletion again.

Crystals can also be used to heal "holes" in the aura. Holes arise from a variety of causes such as physical scarring from an operation, depletion by illness, emotional or mental pain. When you slowly run a crystal over the aura, a hole will feel cold and lacking in energy. Leave the crystal over the hole for a few moments and it will become warm and energized once more. The hole has been repaired. Amethyst, Carnelian, Citrine, and Quartz repair the aura, while Selenite and Kunzite detach unpleasant thoughts and other mental influences.

ABOVE *Wearing the right crystal can energize or protect your aura and your chakras. Here Black Tourmaline is worn to deflect negative energies.*

CRYSTALS FOR THE AURA

AMBER	*An ancient protector. It aligns the aura with the physical body, mind, and spirit. It draws off negative energy and so cleans the aura.*
AMETHYST	*Gently cleanses the aura, heals holes, and protects it, drawing in divine energy.*
APACHE TEAR (CLEAR BLACK OBSIDIAN)	*Gently protects the aura from absorbing negative energies.*
BLACK JADE	*Guards the aura against negativity.*
BLOODSTONE	*Etheric cleanser that greatly benefits the aura.*
CITRINE	*Cleanses and aligns the aura, filling in gaps.*
FLUORITE AND TOURMALINE	*Provide a psychic shield.*
GREEN TOURMALINE	*Heals holes in the aura.*
JET	*Protects the aura against other people's negative thoughts.*
LABRADORITE	*Prevents energy leakage. It protects by aligning to spiritual energy.*
MAGNETITE	*Strengthens the aura.*
QUARTZ	*Cleanses, protects, and increases the auric field, sealing any holes.*
KUNZITE AND SELENITE	*Detach mental influences from the aura.*
SMOKY QUARTZ	*Grounds energy and dissolves negative patterns encased in the aura.*

AMBER

AMETHYST

APACHE TEAR

BLOODSTONE

CITRINE

TOURMALINE

LABRADORITE

KUNZITE

GREEN TOURMALINE

QUARTZ

MAGNETITE

JET

SMOKY QUARTZ

SELENITE

AURIC DEPLETION

The photograph at the top shows an aura with "holes." The person was depleted by physical illness, the death of a close friend, and ill-wishes from a business associate. The color photograph on the right shows how disturbed the aura was. Holding a Clear Quartz crystal sideways across her body, at the navel, extended the aura considerably and repaired virtually all the holes, as shown in the lower picture.

PSYCHIC ATTACK

Psychic attack happens when someone, deliberately or inadvertently, sends thoughts to you that harm you or wishes you ill. This can arise out of jealousy, envy, anger, or ignorance. A Black Tourmaline crystal worn around your neck deflects the energy harmlessly away. A Fire Agate reflects the attack back to its source so that the source can understand the effects of such thoughts. Rose Quartz replaces aggression with love.

Crystals for personal protection

Celestite Attracts a guardian angel.
Amethyst Connects to divine protection.
Black Tourmaline Repels negative energy and psychic attack, and creates a protective shield.
Pyrolusite Repels negative energy from the aura; dispels interference from the physical or spiritual world.
Fire Agate Reflects harm back to its source so that the source can understand its effect.

ABOVE *This aura photograph shows almost total energy depletion. Angelic protection had been invoked to prevent complete energy loss.*

MENTAL INFLUENCE

There may be people around who feel strongly that they know what is best for you or what you should do. The intent may be kindly and well meaning, but it can be manipulative. You may pick up their thoughts and act on them, unaware of the source, and their influence can be subtle and yet powerful. Fortunately crystals can dispel mental influence.

Kunzite and Selenite both help to dispel mental influence from your aura. The easiest way to deal with it is to "comb" the aura with the crystal, working from the top of the head down to the feet.

FIRE AGATE

PYROLUSITE

Talismans and amulets

Talismans and amulets have been used for protection since time immemorial. Made from precious or semiprecious stones, they are imbued with a magical, protective force and, in many cultures, are believed to carry a divine essence.

RIGHT *The Eye of Horus, an ancient Egyptian protective symbol, appears here with the figure of eternity. This gold amulet incorporates many talismanic stones, primarily Carnelian. It came from the tomb of Tutankhamun.*

MAGICAL PROTECTION

In Egypt, gold, silver, and precious stones were used for amulets and talismans because they were considered to be the flesh of the gods themselves. Statues and amulets were imbued with the essence of the god or goddess – and with their attributes. They were believed to be alive. Statues were made both for group protection and for personal use. Amulets were carved out of precious stones such as Lapis Lazuli or Turquoise.

Less exalted, but considered even more effective, were the semiprecious gems Chrysocolla, Malachite, Gypsum (Selenite), Hematite, Calcite, and Magnetite. Even the most humble Egyptian peasant wore an amulet around the neck, and Egyptian aristocracy were buried with fabulous talismans to ensure safe passage to the Other World. Egyptians today still wear the protective Eye of Horus or the scarab, and vendors on every street corner press a scarab onto the tourists "for luck."

RIGHT *The scarab beetle was imbued by the Egyptians with protective energy. It was secreted within the mummy wrapping to protect the soul on its way to the Other World, and is still offered to tourists for protection today. This scarab bracelet, from the tomb of Tutankhamun, features gold, Lapis Lazuli, Carnelian, and Turquoise.*

The belief in the power of talismans and amulets was strong in the Americas, the East, and the West. From ancient Egypt it passed into the ancient Greek and Roman worlds and then carried over into medieval times. In Europe in the Middle Ages, for example, peasant and lord alike carried an Amethyst amulet to protect against drunkenness. The wearing of gemstone rings stems from practices such as this. Crystals are today still believed to have powers of protection.

AQUAMARINE

AMULETS IN COMMON USE

AQUAMARINE *Gives protection against drowning, counteracts the lure of spirits of darkness, and procures wise spirits.*

MOONSTONE *Protects against the perils of travel, and brings good fortune.*

TURQUOISE *Used on horses' bridles to make them more sure-footed.*

CARNELIAN *Protects from the powers of the devil. Also protects from injury by falling walls or buildings.*

BLOODSTONE *Gives courage and protection against kings and despots. Said to preserve the bodily health of the wearer and guard against deception.*

CHRYSOLITE *Set in gold it dispels the terrors of the night and drives away evil spirits.*

MOONSTONE

JADE *Protects children against childhood diseases, and brings prosperity to their elders.*

JASPER *Protects from the bite of venomous creatures.*

AMETHYST *Guards against drunkenness, keeps one safe.*

AMBER *Highly prized for its talismanic powers and protective function.*

DIAMOND *A stone of good fortune. Believed to offer protection against plague and pestilence. Brings victory to the wearer, courage and strength. Hildegard of Bingen believed a Diamond to be an enemy of the devil. Talismanic power works when given as a gift and when the stone touches the skin.*

EMERALD *Renders the power of a magician void.*

JET *One of the earliest talismans, set in silver to draw off negative energy.*

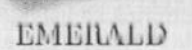

EMERALD

ABOVE *The Chinese have used Jade as an amulet for thousands of years because it is believed to bring good fortune. This amulet incorporates many colors of Jade.*

CREATING YOUR OWN AMULET

Choose a crystal that has the protective properties you need. Hold it in your hands for a few moments and visualize it bringing those qualities of protection to you. Either wear it around your neck or keep it with you wherever you go.

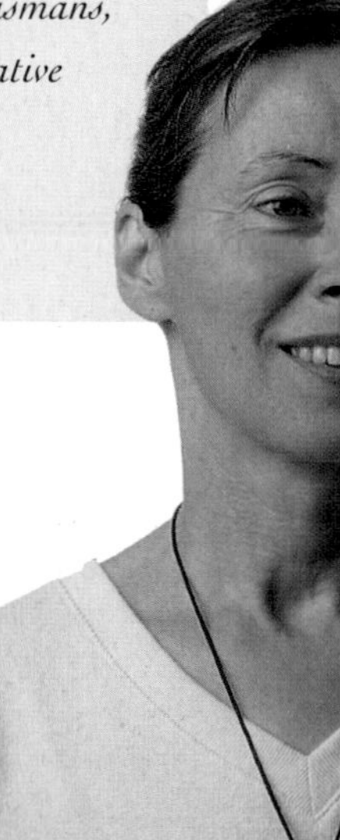

5 Crystal Divination

Scrying is the art of looking into a crystal to divine the future. Traditionally, certain crystals had a divinatory meaning attributed to them. For example, Bloodstone was a weather oracle. When the stone was placed in water in the rays of the sun, it created a reddish tinge. It was said to have the power to turn the sun blood-red, creating extreme weather conditions. The ancient chronicler Damigeron claimed it produced rain and "audible oracles" to announce the future.

Polished spheres of Beryl and rock crystal have been used for millennia. Medicine men and shamans the world over used these crystals to divine the future.

The Druids and Celts regarded Beryl as the best stone for scrying and it is still known as a "stone of power." Merlin, that magical figure from Arthurian myth, inherited Druidic power and carried a crystal ball about his person.

Crystal balls have been used for centuries to gaze into the future. Merlin would have carried a crystal ball much like one of these.

Crystal balls were popular in Europe in the Middle Ages and were passed down from master to apprentice. Queen Elizabeth I of England called upon the services of a master scryer, the mysterious Dr. John Dee. A mathematician, astrologer, and alchemist, Dr. Dee had been imprisoned by Elizabeth's sister Mary Tudor as a magician, notwithstanding that he was Mary's astrologer.

The great Renaissance physician Paracelsus wrote that "conjuring crystals" involved "observing everything rightly, learning and understanding what was."

For hundreds of years crystal balls were chipped by hand in Japan. In France, Germany, and the United States, they were turned on a grindstone and patiently polished. Demand for the services of a crystal ball was high.

Crystal balls

Peering into a crystal ball focuses intuition, allowing the rational mind to slip out of gear. Light refracting on the crystal's surface holds the eye, fixing the optic nerve. With the eyes fixed, the crystal ball mists over. Within the mist, images form. A seer reads these pictures or a crystal "spread," in a manner similar to turning the pages of a book. The images come from within the mind of the reader, but they have been externalized in the crystal for clearer sight. You can glance sideways into an Apophyllite crystal for the same effect.

ABOVE *Scrying was not always frowned upon by the Church. This German plate of the 1700s shows a monk gazing into his crystal ball.*

CHOOSING A CRYSTAL BALL

A crystal ball is a very personal thing. By tradition, it is better to have one given to you, but you can purchase one. Crystal and metaphysical shops, psychic festivals, and New Age fairs are useful sources. Most crystal balls are clear in color, although Quartz ones may have planes or flaws within. Smoky Quartz has traditionally been used for crystal gazing, but Amethyst, Beryl, Selenite, Obsidian, and other crystals are also used.

ABOVE RIGHT *When choosing your crystal ball, weigh up several to find exactly the right one for you.*

ABOVE *This Amethyst crystal ball has wonderful planes and flaws within it that refract light, helping the eyes to go out of focus. The energy of the crystal calms and relaxes the mind, and heightens the intuition.*

Before you purchase a crystal ball, weigh several in your hands. Look at the size and consider how it feels to you. They may be slightly different shapes: which one feels most comfortable to hold? When you have chosen your ball, make sure you cleanse it properly before use. Do not let other people handle it.

LEFT *It is usual to store a crystal ball wrapped in a cloth to protect it.*

CRYSTALS FOR ENHANCING SCRYING ABILITY

- *Amethyst*
- *Apophyllite*
- *Azurite with Malachite*
- *Celestite*
- *Chrysocolla*
- *Lapis Lazuli*
- *Selenite*

LEFT *Sit quietly and focus your energies before you begin to read your crystal ball. A Selenite crystal like this one is particularly useful for understanding the past.*

USING A CRYSTAL BALL

Hold the ball in your hands for a few moments to charge it up. If you have a particular problem, focus on it and consider possible solutions. Then place the crystal on its stand on a black silk or velvet cloth. Some people like the room to be dim, others have one bright source of light. You may find that candlelight is favorable for reading.

Let yourself relax and focus on your crystal. Do not try to see images, let them form naturally. As your eyes fixate, the ball will go out of focus. It may appear to cloud from within as though a mist is forming in the crystal. Keep your eyes relaxed and watch for symbols or pictures appearing either in the crystal or in your mind's eye. When the mist clears, make detailed notes of what you see, and the impressions you receive. Some of it may seem to be nonsense, like the images you see before going to sleep. If you are patient and persevere, you will come to understand the meaning.

When you are ready to try out your skills on someone else, ask him or her to cup hands around the ball, but not to touch it. After a few minutes the crystal will be attuned to your sitter's energies. Then proceed as usual. Ask for feedback because this is the way you will learn.

BELOW *Make a note of what you see. This way you will start to build up your own interpretation dictionary for the symbols you see.*

Crystal dreams

Dreaming was used by the ancients as a means of seeing what was to come. To this end, dreams were deliberately induced, and great emphasis was placed upon what they revealed. Not only can crystals play their part by inducing dreams, but should crystals themselves appear in dreams, they add to the interpretation.

OPAL

RUBY

LAPIS LAZULI

ONYX

In dreamlore, if you dream of Opals you will receive great possessions; if you see an Amethyst, peace of mind will come through unexpected good news; and an Aquamarine foretells a happy love life – unless you dreamed that you lost it. Polished Agate warns against being drawn into arguments between friends; Bloodstone portends an unhappy love affair; but Ruby signifies passion. Diamonds have a double meaning. If you own them in life, they signify slight losses. If you do not, dreaming of them signifies some profit but not as large as you hoped. Jade, however, signifies prosperity, and Emerald good fortune. Jet warns of sad news, and Quartz that you may be cheated by someone you trust, but Lapis Lazuli brings happy adjustment to circumstances. To see Onyx indicates you are holding back on an important decision and need advice. To break this brittle stone in your dream signifies luck. Sapphire is another dual stone. To see someone else wearing it indicates a change in social status through influential friends, but if you wear it, beware impulsive behavior.

To see light shining through a crystal points to a speedy resolution of a problem. Jewels hidden in a cave and guarded by a serpent represent spiritual treasure in the unconscious mind.

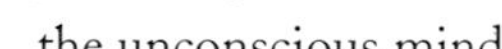

RIGHT *Crystals have traditionally been used to induce dreams. However, they may also appear in your dreams.*

Crystal layouts

A handful of tumbled crystals can be used for crystal divination. Put them into a bag, ask your question, and take out the first two or three that your fingers touch. This will give you your answer.

You can also use a purpose-made board onto which you throw your crystals. You then combine the answers on the board with the wisdom of the crystals.

Precious gems were often used for divination. If you do not have precious gems, substitute Clear Quartz for Diamond, Peridot for Emerald, Garnet for Ruby, and so on.

DIVINATORY MEANINGS

AGATE *Worldly success*

AMETHYST *Life changes and shift in consciousness*

BLACK AGATE *Courage and prosperity*

RED AGATE *Health and longevity*

AVENTURINE *Growth and expansion*

BLUE LACE AGATE *Healing needed*

CITRINE *Celestial wisdom*

DIAMOND *Permanence*

EMERALD *Fertility*

HEMATITE *New opportunities*

JADE *Immortality and perfection*

RED JASPER *Earthly affairs*

LAPIS LAZULI *Divine favor*

QUARTZ *Clarify issues*

ROSE QUARTZ *Love and self-healing*

RUBY *Power and passion*

SAPPHIRE *Truth and chastity*

SNOWFLAKE OBSIDIAN *End of challenging time*

TIGER'S EYE *All is not as it seems*

SNOW QUARTZ *Profound changes*

UNAKITE *Compromise and integration*

RED JASPER

BLUE LACE AGATE

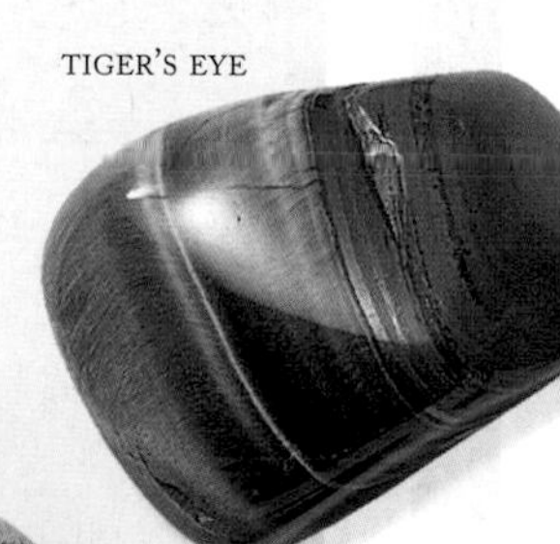

TIGER'S EYE

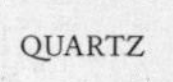

QUARTZ

SNOWFLAKE OBSIDIAN

C116
R144
R141
R133
D105
C118

6 Crystal Qualities

Crystals have very special properties, some of which are shared with all crystals, while others are unique to specific ones. These special qualities can be harnessed and put to use. Certain qualities work on a physical-energy level, others on a more subtle vibration. The power of crystals to attract is particularly useful. Crystals have esoteric as well as physical properties. They produce piezo-electricity when compressed. This means that energy, and sometimes light, is produced when a crystal is squeezed. The minute expansion that follows compression releases electrons and then re-absorbs them, creating energy. Some crystals, such as Topaz, have a static charge when rubbed or heated.

The largest crystal family is Quartz, which is found on every continent. Quartz occurs in large, six-sided, crystals and microscopic ones, singly or in groups, in every possible color. It is made up of the basic building blocks of minerals, silicon, and oxygen. A much-studied crystal, Clear Quartz was used to transmit the first radio waves. Modern technology relies on its ability to receive and transmit energy. It made the computer revolution possible.

Quartz has other, more esoteric, properties. An excellent healing stone, it is an energy regulator for the physical body. It widens the subtle energy field or aura. It enhances muscle testing (using the body's muscles to test whether organs are functioning properly, if substances are beneficial, and much else besides). Quartz can heal, protect, attract wealth, enhance intuition, and disperse negativity.

The particular properties of Quartz mean that it can be electrostatically colored – as shown by this beautiful piece of Aqua Aura, colored with gold. Treating a crystal in this way changes its properties. The gold intensifies its energy, which can be used to cleanse and smooth the aura and activate all the chakras.

Grounding

Brown, black, and green stones ground - and therefore earth - energy. You can ground your subtle bodies into your physical body, anchoring your feet firmly on the floor; bring the spiritual into the physical; or earth negative energies into the ground for transmutation.

Carnelian

A translucent orange-red stone, Carnelian is strongly attuned to earth. This stone makes you feel comfortable within your surroundings – like having an anchor into the physical world. Carnelian is useful for people who meditate since it clears extraneous thoughts. It facilitates direction and control of your life.

Green Fluorite

A transparent, often cubic crystal, Green Fluorite grounds excess energy. It prevents overstimulation on a psychic level, anchoring the intuition into everyday reality while enhancing spiritual perception. Holding a green fluorite wand calms and grounds energy during meditation.

Hematite

Hematite has a metallic luster. Silver when tumbled, in its natural state it ranges between steel gray, grayish-black, and reddish, and may have red spots on the surface. It is an iron oxide, magnetic, and strongly connected to the earth. It aids people who haven't got their feet on the ground. It sorts out issues, making it possible to move ahead. Hematite dissolves negative energies, transforming them into positive ones.

Boji Stone

A metallic stone with a high pyrite content and strong affinity with the earth, there are two "sexes" of Boji Stone. The female is smooth and rounded; the male has square protruberances. Aligning chakras and subtle bodies, Boji Stones gently ground after meditation or healing.

OTHER GROUNDING CRYSTALS

✤ *Amber* ✤ *Brown Jade* ✤ *Jasper* ✤ *Onyx* ✤ *Smoky Quartz* ✤ *Snowflake Obsidian*

BELOW *Crystals help you to be grounded in your body, protecting your energy field. Keeping a Hematite crystal in your pocket earths you.*

HEMATITE

GREEN FLUORITE

MALE BOJI STONE

CARNELIAN

FEMALE BOJI STONE

Dispersing negative energy

SMOKY QUARTZ

Negative energy makes you, your physical body, or your surroundings feel heavy. Cleansing these energies with crystals brings instant lightness. Black stones absorb negative energy while other colors stimulate positive energy.

LEFT *Smoky Quartz is the perfect stone for absorbing negative energy.*

Smoky Quartz
A transparent or translucent stone, Smoky Quartz varies in color from light brown to almost black. It makes a perfect grounding stone, because it is attuned to the earth. It has the ability to neutralize personal and environmental negative energies.

Green Calcite
A translucent, waxy stone, Green Calcite clears negativity out of a stagnant situation, helping adjust to new challenges.

Chalcedony
Chalcedony encourages introspection. It gently releases self-doubt and negative beliefs, creating a new enthusiasm. It has the ability to absorb and dissipate negative energy. A powerful cleanser of energy, it is one of the few crystals that does not always have to be cleaned after use.

Jade
Jade comes in several colors, each of which has specific properties. All colors of Jade share the ability to release negative thoughts and energies.

Jet
A hard, fossilized wood, Jet is a powerful absorber of negative energy. It overcomes unreasonable fears, creating a more powerful outlook.

Apache Tear
The translucent, water-smoothed Apache Tear is an excellent remedy against negative energy from any source, but especially against the negativity of other people. The energy is absorbed and then sent to the "rubbish dimension."

Citrine
A clear yellow to brownish-amber Quartz, Citrine has the ability to dissipate and transmute negative energies without absorbing them, and can be used without cleaning. Citrine dispels negative feelings and aids in accepting constructive criticism.

Amber
Amber is an excellent energy cleanser, drawing negative energy from the body and allowing it to heal itself. It also clears environmental energies.

ABOVE *Dispersing negative energies leaves you positively fizzing with energy and vitality.*

APACHE TEAR

AMBER

Creativity

RIGHT *Choose the right stone for your creative needs. Yellow Fluorite would be an excellent aid to an orchestra member. For a solo performance, Citrine is more appropriate.*

When you use your creative energy you are using your life essence, your intuition, and your intellect in combination. Creativity sparks ideas, bringing inspiration to life. Without creativity there would be no new projects, no innovative solutions, no artwork, literature, or music. Crystals can fire your creative spirit. Yellow and blue-green crystals strongly stimulate creativity.

MALACHITE

CHRYSOPRASE

CITRINE

TURQUOISE

CHIASTOLITE

Chrysolite
Heightens inspiration.

Chrysoprase
Opens to new surroundings, bringing fresh ideas. It draws out unknown talents.

Citrine
Harnesses creative energy. It brings inner calmness, making you more open to constructive criticism.

Green Tourmaline
An excellent problem solver. It rejuvenates and stimulates, helping to avoid negative energies that drag down.

Turquoise
Very beneficial for creative problem-solving. It helps to understand how things work and where you fit into the scheme of things.

Aventurine with Malachite
Clears mental blocks. This combination opens the intuition and attunes to divine inspiration.

Amazonite
Helps to filter intellectual information and combine it with intuition to fire creative processes.

Carnelian
Draws on past experiences to strengthen creativity. It also encourages curiosity and enthusiasm.

Chiastolite
Balances creativity with practicality, putting ideas into practice.

Yellow Fluorite
Supports both the intellectual and the creative processes, and aids cooperation between members of a team.

YELLOW FLUORITE

AMAZONITE

ORANGE CALCITE

GREEN TOURMALINE

CARNELIAN

EMERALD

Vitality

TOPAZ

Lack of vitality results from a lack of balance in the physical body. Without vitality nothing happens. The energy is dead, life is a drag. Crystals rectify imbalances and blockages, allowing energy to flow freely. With vitality, everything is possible. Life sparkles, new ideas take off. Yellow, orange, and red crystals promote vitality, but red can sometimes be too stimulating.

RUBY

Topaz
Recharges and revitalizes. Its color is like the sun bursting through clouds, and it boosts energy with its warm, vital, life-giving force.

Ruby
A passionate stone that amplifies positive energy, creating a restorative pool to draw on.

Apophyllite
Brings energy and light. Natural Apophyllite pyramids stimulate energy.

Chalcedony
Increases physical energy, and endurance.

Dioptase
Balances yin and yang and clears all the chakras, invigorating the physical, emotional, and mental levels of being.

Aventurine
Protects the heart chakras, preventing leaching of energy and thus enhancing vitality.

Agate
An earthy stone that brings physical and emotional balance. Dendritic Agate helps to direct energy during discordant or disorientating situations, lessening the energy drain.

APOPHYLLITE

Bloodstone
Aids staying in the present moment and renews the physical, emotional, and mental bodies, thereby increasing stamina. It assists the heart to stay open during difficult experiences, strengthening vitality.

CHALCEDONY

Emerald
Creates harmony in the heart, from which vitality is said to flow. It enhances joy and the positive qualities of life.

DIOPTASE

OTHER STONES FOR VITALITY
•Citrine •Garnet•Jasper •Orange Calcite •Rutilated Quartz •Topaz

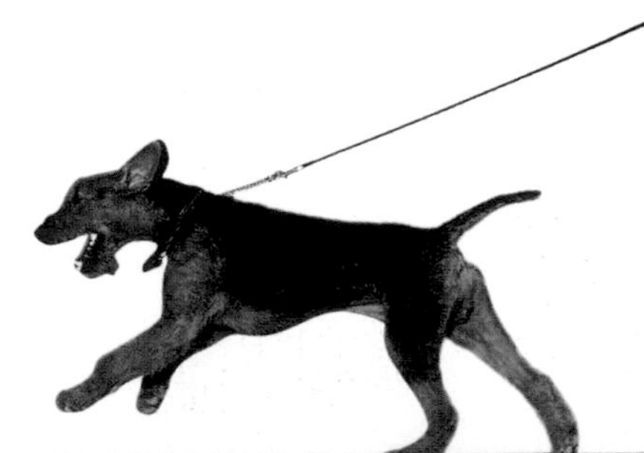
RIGHT *The appropriate crystal soon gets your energy flowing.*

BLOODSTONE

DENDRITIC AGATE

AVENTURINE

Peace and relaxation

BELOW *Carefully chosen crystals, placed on the body, promote peace and relaxation.*

One of the greatest benefits you can receive from a crystal is a deep sense of peace and relaxation. If you feel "out of sorts," a crystal will gently bring you back into balance. If you feel uptight, it will unwind you. If you need to be calmer, a crystal quiets body, mind, and emotions. Placing appropriate crystals on chakras releases energy held there, helping you to be more relaxed.

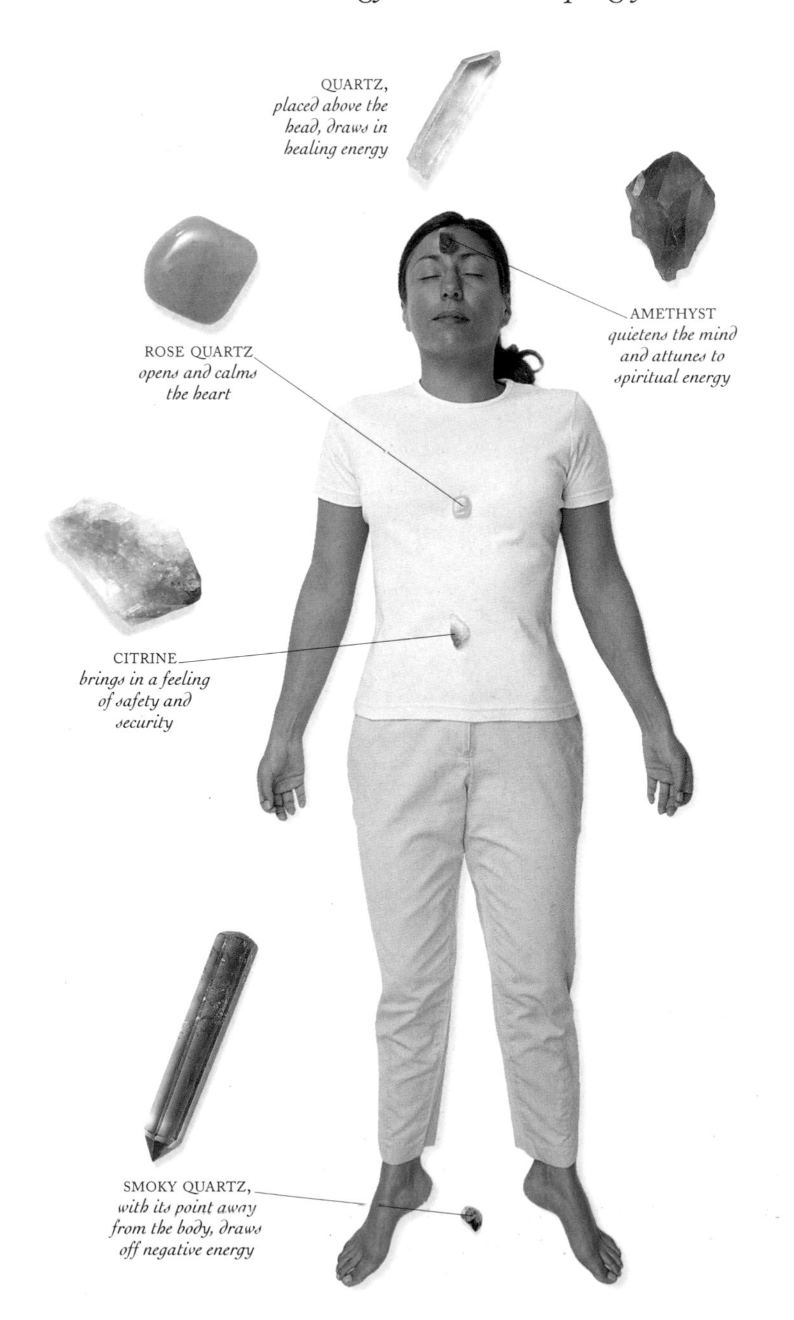

QUARTZ, *placed above the head, draws in healing energy*

AMETHYST *quietens the mind and attunes to spiritual energy*

ROSE QUARTZ *opens and calms the heart*

CITRINE *brings in a feeling of safety and security*

SMOKY QUARTZ, *with its point away from the body, draws off negative energy*

Relaxing crystals can be chosen by qualities or color. Pink and green crystals are particularly calming but some of the violet stones elevate consciousness, raising up out of tension into peace, while some of the yellow stones bring peace of mind.

Quartz

Quartz is a particularly useful crystal for relaxation. Rose Quartz calms and opens the heart, a Smoky Quartz with its point toward the earth draws off stress and negative energy from the base of the spine, while a Clear Quartz on or above your crown draws in healing energies. Amethyst is also a Quartz. You can place an Amethyst point on your forehead to quiet your thoughts. Citrine, yet another Quartz, pointing down from your navel adds to your feeling of safety and security. Lie on your bed with the stones in place for five minutes or so and you will feel the benefit; take 20 minutes and you will feel wonderful.

DIOPTASE

Chrysoprase
The color of green apples, Chrysoprase radiates peace and calmness. Its ability to bring greater insights and to aid you in opening to new situations and surroundings helps you to remain relaxed at times of change.

Blue Jade
Blue Jade is particularly useful for people who feel their situation is beyond their control. It brings patience and an inner serenity, ensuring slow but steady progress.

Kunzite
With its delicate striations, Kunzite lifts the spirits. It is another of the stones that radiate peace. Place it on the heart chakra for maximum effect.

KUNZITE

BLUE JADE

CHRYSOPRASE

RHODONITE

OTHER RELAXING STONES

Amber ✤ Aventurine ✤ Dioptase ✤ Jasper ✤ Peridot ✤ Variscite

Rhodonite
The elegant, softly pink Rhodonite is a particularly useful stone for remaining calm and centered in the middle of trauma or crisis.

Herkimer Diamond
Crystal-clear Herkimer Diamond draws off environmental stress. If you feel particularly stressed out, with your nerve jangling and an inability to sleep, a few of these crystals around your bed will create a safe, calm space in which to rest.

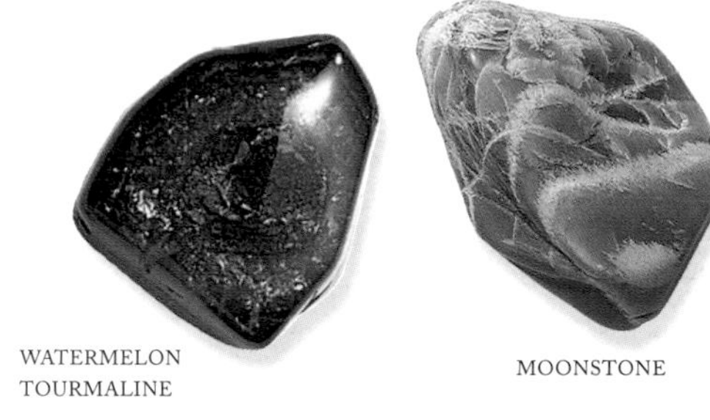

WATERMELON TOURMALINE

MOONSTONE

HERKIMER DIAMOND

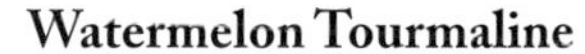

Watermelon Tourmaline
A relaxing combination of pink enfolded within green, Watermelon Tourmaline brings peace through understanding. It aids in making peace with the past and finding a way to resolve conflicts.

ONYX

Onyx
An earthy stone, Onyx brings strength and peace of mind. It is a useful stone for anyone under severe emotional or mental stress, alleviating excessive fears and worries.

Moonstone
Moonstone calms and balances emotional overreactions. It reminds us that all things are part of an ongoing cycle and enables us to receive direction from the unconscious mind.

Confidence and self-worth

ABOVE *You can enhance your self-confidence with a well-chosen crystal.*

With plenty of confidence, you have a natural feeling of self-worth. Valuing yourself gives you confidence, which can overcome inherent shyness and natural reticence. Knowing your own value can help you to stand up for yourself. If you lack confidence, you can easily stimulate this with a crystal. If you are not in touch with your own worth, a crystal will soon introduce you.

Chalcedony
Clears self-doubts. Its ability to aid objective retrospection links you into your own value, which you can then demonstrate to other people.

Chrysocolla
Aids personal confidence and speaking the truth.

Hematite
Bolsters low self-esteem. In ancient times, it was believed to confer invincibility. Nowadays it helps to recognize the truth of your own worth.

Tiger's Eye
Assists you to recognize your inner resources. Its ability to ground endeavors enables confidence to develop.

Rutilated Quartz
Brings resolve and strength of will, self-reliance and the ability to make decisions. It helps you to be quietly confident.

Green Tourmaline
Has the useful property of enabling you to recognize and avoid negative situations before you become entangled in them. This can greatly strengthen confidence.

Azurite
A dynamic stone that gives you the objectivity to examine your own inner workings, and to recognize your own resources.

CHALCEDONY

CHRYSOCOLLA

Meaning and purpose

If you know where you are going and what you are meant to be doing, you are stimulated and energized. If you understand the meaning of life, you can surmount obstacles. Without meaning, life is without direction and dull. With a sense of who you are and what you are here for, life is purposeful.

THE LIFE PATH

In crystal lore it is recognized that we all have a life path, a course that we must follow if we are to feel fulfilled and that is set by the soul not the ego. Crystals such as Selenite and Quartz can link you into this path. A life-path crystal is a long, thin, Clear Quartz crystal with one or more absolutely smooth sides. Helping you go with the flow and follow your destiny, it leads you to joy.

Sodalite

Aids objectivity and assessment of goals. It helps you to simplify your life, identifying the path you need to walk for maximum fulfillment.

CARNELIAN

SODALITE

QUARTZ

TOPAZ

Carnelian

The stone to use if your life has fallen into a rut. It stimulates a new train of thought, increasing motivation and opening new avenues for success. It aligns you with your life purpose.

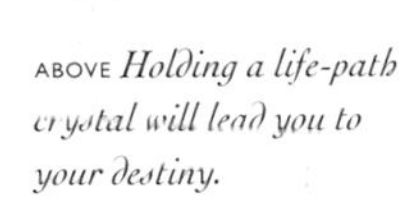

ABOVE *Holding a life-path crystal will lead you to your destiny.*

Quartz

This crystal provides an optimum state of openness. It guides you in the search for purpose and meaning in your life.

Topaz

Attunes you to the unexpected and to limitless possibilities. It sheds its golden light on your path, helping you recognize your life's purpose.

MAHOGANY OBSIDIAN

Selenite

Links you to angelic guidance and teaches about higher meaning and purpose. It unites you with your soul's purpose.

Mahogany Obsidian

Revitalizes your life purpose.

SELENITE

Opening the intuition

Intuition is like a sixth sense. It sees beyond the confines of the intellect to what cannot be known by normal thought processes. Intuition makes leaps into the unknown and comes up with answers. It is an inner awareness that, with the help of crystals, can be honed and developed.

CELESTITE

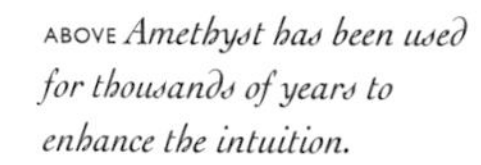

ABOVE *Amethyst has been used for thousands of years to enhance the intuition.*

SODALITE

Amethyst
Strengthens intuition, centering the mind so that it is receptive to answers that rise up into awareness.

Yellow Calcite
Aids meditation and focuses the intuition.

Chrysolite
Makes you more receptive, heightening psychic experiences.

Lapis Lazuli
The sacred stone of ancient Egypt, it focuses the intuitive process and opens the psychic centers of the mind.

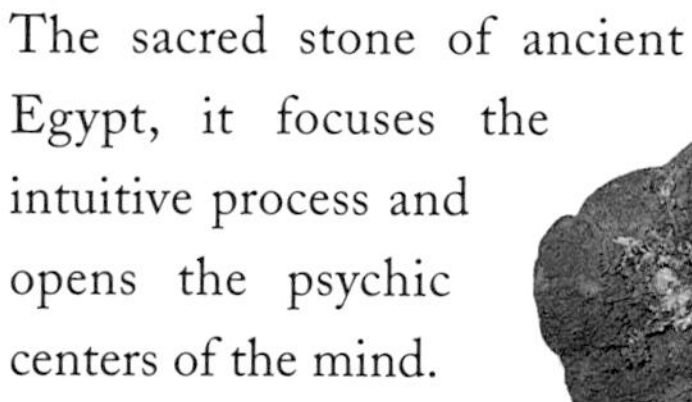

LAPIS LAZULI

Moonstone
This stone is linked to the moon, the traditional repository of knowledge from the subconscious mind and the intuition.

MOONSTONE

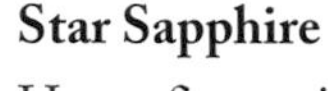

Star Sapphire
Has a five-pointed star within it that draws you into its depths, opening the intuition along the way.

Apophyllite
Releases intuition when held to the third eye.

Amazonite
Unlocks psychic vision when held to the third eye.

AMAZONITE

Selenite
Lifts you onto a higher plane of consciousness, giving effortless access to the intuition.

Celestite
Helps you to contact the intuitive guidance of the angelic realm.

Sodalite
Imparts discernment and objectivity to intuition.

Smoky Quartz
A traditional stone for crystal gazing that focuses and grounds intuition.

SMOKY QUARTZ

Azurite
Opens the psychic faculties.

AZURITE

SELENITE

APOPHYLLITE

Past-life recall

GARNET

HERKIMER DIAMOND

Past lives can be glimpsed through dreams, intuitions, spontaneous recall, and meditation. When the soul sees that the time is right for the conscious personality to look at past lives, it brings them to the surface. You can place a suitable crystal, such as Variscite, under your pillow and ask for a dream to reveal a past life.

CARNELIAN

With the right guidance – **and this is not something that should be done alone** – past-life recall through hypnosis and other techniques can access previous wisdom, and it can release blockages that have their roots in the past.

Raising consciousness

Azurite enhances spiritual awareness and promotes imagery. Meditating with an Azurite crystal on the third eye opens the inner eye of the soul, releasing memories. It facilitates letting go of the past.

Mind-expanding Selenite accesses a different state of consciousness. Rubbing the stone with a finger or gazing into the milky depths of a Selenite crystal ball stimulates images of the past.

Variscite, which looks a bit like Turquoise, aids dream recall and stimulates past life memories, as does Garnet.

Old memory

Cuprite can be helpful in pinning down a repeating pattern of difficulties with authority or father-figures.

Serpentine has a very old memory. With its ability to focus on the spiritual reasons for existence, it takes you back into the far past to retrieve ancient wisdom and knowledge of your past lives. As with all brown stones, Brown Jasper has a strong connection to the past. It aids past-life regression, revealing hidden stumbling blocks carried from life to life. Brown Jasper itself does not release and heal blocks; other crystals are needed for this.

Herkimer Diamonds hold a store of ecological memory, and they facilitate understanding of what has been seen. They have the ability to release past-life blockages. The access Carnelian gives you may not be to your own past lives but to historical events. The insights this provides help you to open to new possibilities in the present life.

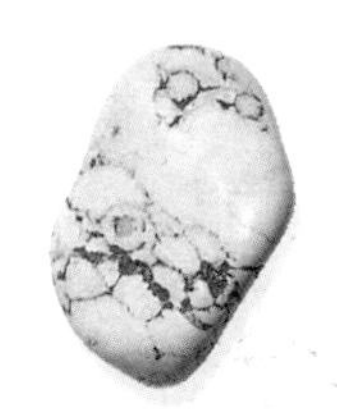
VARISCITE

ABOVE AND LEFT *Placed on the third eye, Variscite or another stone such as Amethyst can facilitate past-life recall, but this should not be practiced alone.*

Abundance

Abundance is much more than wealth. Abundance is being connected to the source of all that is, manifesting a constant flow to support, nourish, and expand life, love, and creativity. It is unlimited, effortless, and joyous. Crystals connect you to abundance, enabling you to trust that what you need will come – in abundance.

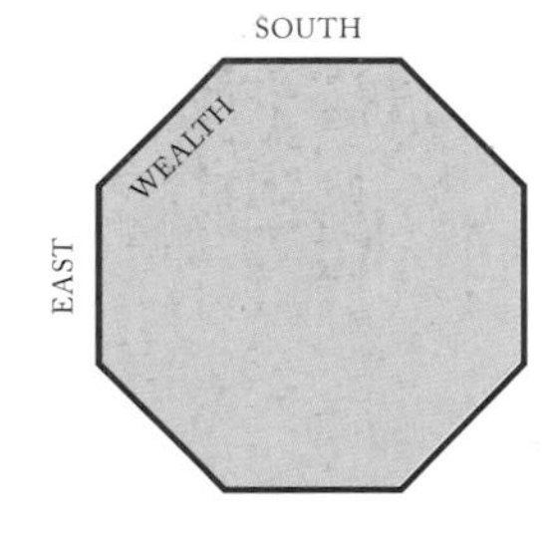

ABOVE *In one system of Chinese Feng Shui, the four directions of the compass are used (with south where we would place north) and the wealth corner is always in the southeast section.*

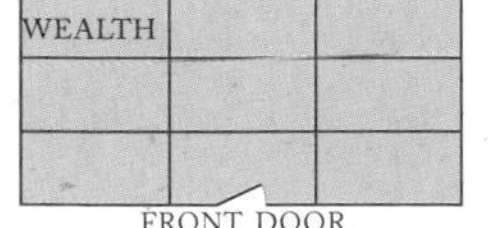

ABOVE *In another form of Feng Shui, the wealth corner is the far left of the front door.*

THE WEALTH CORNER

According to Chinese Feng Shui, different parts of a house relate to different aspects of life. The wealth corner is the corner farthest away and to the far left of the front door (or the southeast corner, depending on which system is used). If you place an abundance crystal in the wealth corner of your house, it attracts riches. If your bathroom is located here, keep the door shut and the toilet seat down, otherwise, no matter how much wealth is generated, it will "go down the toilet" and be lost.

Yellow Sapphire

Associated with Hindu god of prosperity, Ganesh, Yellow Sapphire attracts wealth to home. If worn, it should actually touch the finger.

Abundance Crystal

An abundance crystal – one long Quartz crystal with many tiny crystals at the base – encourages dreams and promotes well-being and love. It brings blessings and growth on all levels.

Tiger's Eye

The beautifully banded Tiger's Eye is used for jewelry. Attracting helpful people and material things, it points out the most advantageous way to do things, naturally attracting abundance. It brings an awareness of one's needs, balancing these against those of another. It stimulates the acquisition of wealth and creates the stability needed to maintain wealth.

Hawk's Eye

A darker variation of Tiger's Eye, Hawk's Eye can look exactly like the piercing eye of a hawk. It is stimulating and energizing and, like Tiger's Eye, draws helpful people and material things into your life.

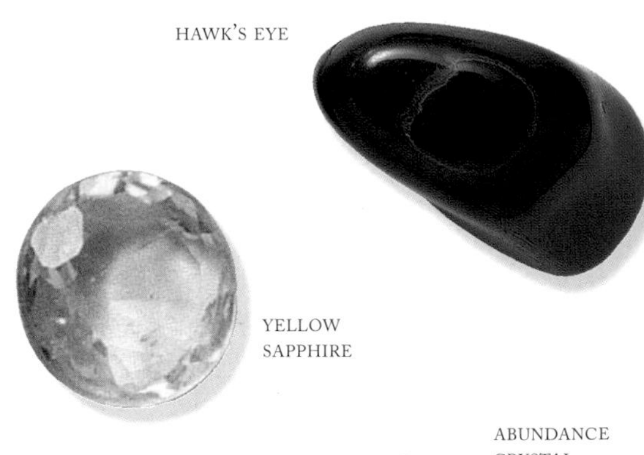

HAWK'S EYE

YELLOW SAPPHIRE

TIGER'S EYE

ABUNDANCE CRYSTAL

PERIDOT

Topaz

Topaz is like having the warmth of the sun shine upon you. It helps you to appreciate the joy of life and to expand, making room for abundance. Topaz taps into your own natural resources, opening unlimited possibilities. It unlocks your philanthropic energies, teaching that when you give, you open yourself to receive. Topaz set in a silver wand is a powerful way to manifest abundance energies. Topaz is unusual in that its facets and points carry both negative and positive charges. These alternating currents link to the forces that manifest desire.

TOPAZ

Carnelian

Carnelian is an excellent stone for helping oneself, and it is said that God helps those who help themselves. Carnelian improves motivation, getting you out of a rut, leading to enriching experiences.

CARNELIAN

Peridot

Peridot brings things quickly, but it may be hard to handle. It works best for someone who is already in control of life and knows where he or she is going.

Dendritic Agate

A "stone of plenitude," Dendritic Agate was used in ancient Greece to ensure plentiful crops. It enhances yields of any kind, giving abundance and fullness of life, encouraging joy in the moment.

Citrine

Citrine was traditionally used by Indian merchants to attract more wealth to their cash box. Place it in your wealth corner to attract abundance.

ABOVE *The Hindu god of prosperity, Ganesh, is associated with yellow crystals. Here, he has been carved from Tiger's Eye.*

CITRINE

ADDITIONAL STONES FOR ABUNDANCE

Diamond ✤ Bloodstone ✤ Moonstone ✤ Moss Agate ✤ Red Garnet ✤ Ruby

7 Crystal Astrology

Crystals have always had a close link with astrology. Birthstones ground stellar influences, and crystal properties are amplified by the Zodiac and the planets that inhabit it. Each birth sign not only has one or two birthstones that resonate strongly with it but also has a plethora of associated crystals. This is because a Zodiac sign bridges two months, each month having its own crystal correspondences, and signs are also "ruled" by planets with their own crystal affinities. Zodiac crystals may therefore be influenced by planetary rulers or by the month of birth, as well as by a crystal's astrological affinity.

Birthstones have been used for thousands of years to bring celestial influences down to the earth.

Astrologers originally worked with five planets visible to the naked eye, plus the luminaries – the Sun and the Moon (also known as "planets" in astrological terminology). The Sun and the Moon each ruled one sign, and the other planets ruled two. Each "planet" had crystal and metal affinities that amplified its power. Nowadays, there are at least ten astrological planets. As new planets are detected, so metals and crystals are discovered that resonate with them. Chiron, for example, wandered into our solar system eons ago but was only recently recognized. A maverick body, Chiron has a powerful healing influence and resonates with Virgo, Scorpio, and Sagittarius (rulership, however, has not yet been assigned). The newly discovered Charoite stone shares Chiron's qualities of healing, synthesis, integration, and transmutation.

The Zodiac is divided into four elements: Earth, Water, Air, and Fire. Each element has a different function and characteristics. Signs that share the same element resonate together and may share crystal affinities.

Earth crystals

Earth shapes matter. Earth is the element of the material world, it is stable and grounded. Practical and productive, it is concerned with self-sufficiency and security. Earth signs are sensual signs. This is the element most involved in the physical body and the senses. The Earth signs are Taurus, Virgo, and Capricorn.

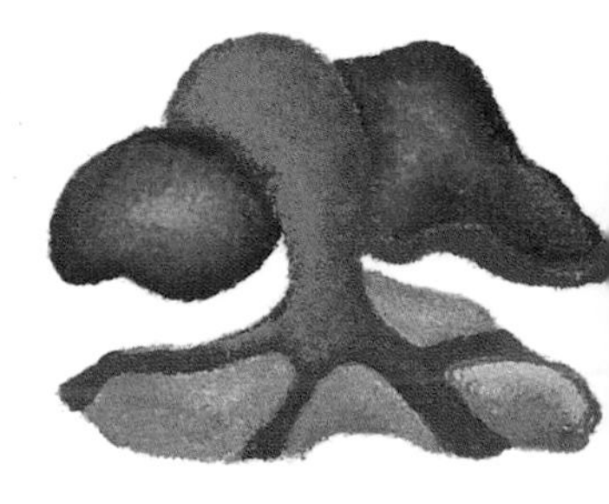

RIGHT *Material and sensual Earth signs like to enjoy the pleasures of life.*

TAURUS

April 20 to May 20

RULING PLANET: VENUS

BIRTHSTONES: EMERALD, TOPAZ

Taurus is a strong, dependable sign. Its keynotes are structure and endurance, and many of the crystals associated with Taurus ground energy into the physical body. Kunzite aids Taurus in removing obstacles that would otherwise overly engage this sign's patient persistence. It also connects Taurus into deep centeredness and inner security. However, as the Emerald birthstone shows, Taurus is also a hedonistic sign. It is, after all, ruled by voluptuous Venus. Emerald is one of her stones: a symbol of successful love.

EMERALD

TOPAZ

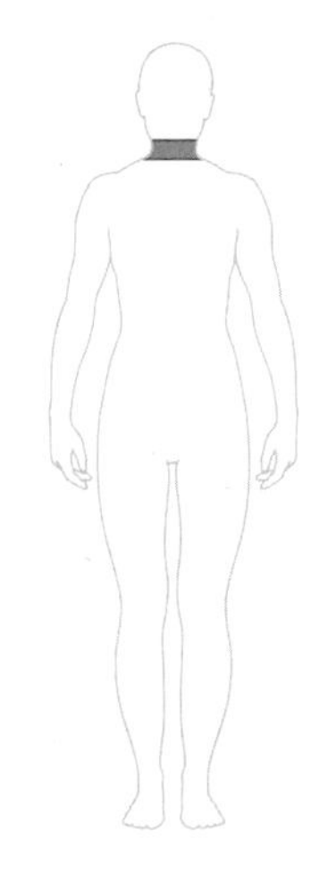

ABOVE *In medical astrology, Taurus rules the throat. Aquamarine crystals can soothe a sore throat or help open the throat chakra.*

AFFINITIES

Aquamarine, Azurite, Boji Stone, Diamond, Emerald, Kunzite, Lapis Lazuli, Malachite, Rose Quartz, Rhodonite, Sapphire, Selenite, Tiger's Eye, Topaz, Tourmaline, Variscite

VIRGO

August 23 to September 22

RULING PLANET: MERCURY

BIRTHSTONES: PERIDOT, SARDONYX

Virgo is a practical and well-organized sign. Its keynotes are efficiency and service. As befits its ruler, Mercury, it is analytical and discriminating. Peridot, one of the birthstones for Virgo, magnifies the inner aspect of situations – enabling Virgo to get to the nub of the problem or experience.

Virgo seeks perfection and Amber aids connection of the self with universal perfection. Chrysocolla attunes Virgo to universal harmony, leading to a perfect state of health. Health – and the lack of it – is a peculiarly Virgoan obsession and this sign seeks to understand the psychosomatic nature of disease.

Above all, the purity of crystals delights the chaste Virgoan heart.

SARDONYX

PERIDOT

AFFINITIES

Amazonite, Amber, Blue Topaz, Dioptase, Carnelian, Chrysocolla, Citrine, Garnet, Magnetite, Moonstone, Moss Agate, Opal, Peridot, Rutilated Quartz, Sapphire, Sardonyx, Sodalite, Sugilite

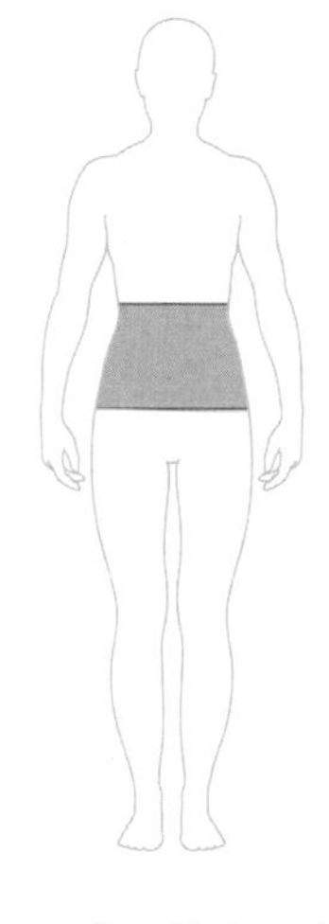

ABOVE *In medical astrology, Virgo is associated with the nervous system and the intestines.*

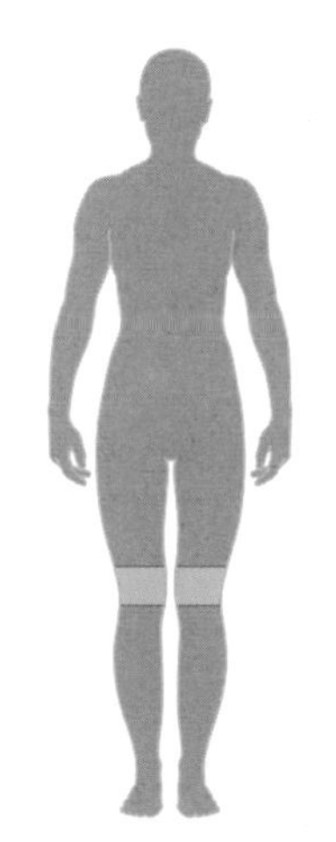

ABOVE *In medical astrology, Capricorn rules the skeletal system and knees. It is also linked to the skin, the outer limit of the body.*

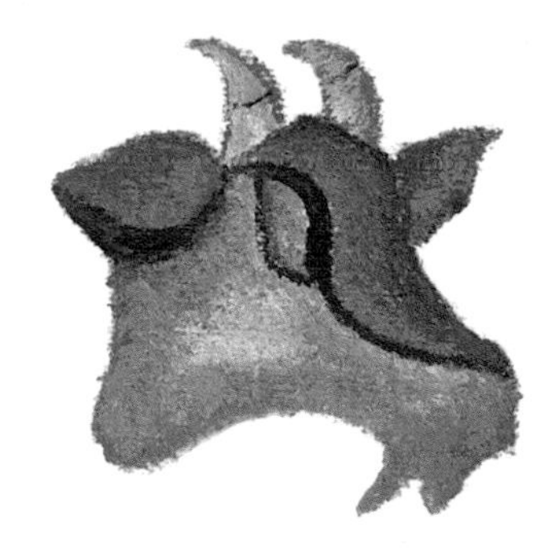

CAPRICORN

December 22 to January 19

RULING PLANET: SATURN

BIRTHSTONES: JET, ONYX

Ambitious Capricorn is a hard-working, dependable sign: a doer. Its keynotes are structure and discipline. This is the sign of the father and authority figures. The seriousness of the Capricorn nature is supported by its birthstone, Onyx, which strengthens resolve and aids self-confidence. Jet is one of Saturn's stones. It is useful in dispelling the Saturnine gloom that so often afflicts Capricorn. Labradorite, with its iridescent flashes of light, lightens the spirit and helps Capricorn understand a strong sense of duty and destiny. It affords a glimpse of spiritual possibilities open to an evolved Capricorn soul.

For Capricorn joints, which can become stiff and rigid, Azurite brings flexibility.

ONYX

JET

AFFINITIES

Amber, Azurite, Carnelian, Fluorite, Garnet, Green and Black Tourmaline, Jet, Labradorite, Magnetite, Malachite, Onyx, Peridot, Quartz, Ruby, Smoky Quartz, Turquoise

Water crystals

Water shapes intuition and emotional experience, and is linked to the feeling and emotional world. The element of intuitive perception, Water signs are inward-looking and sensitive. This is a passive, rhythmic element that ebbs and flows. Water signs need time to withdraw, to process emotional experiences, and then to reemerge into the world. The Water signs are Cancer, Scorpio, and Pisces.

RIGHT *The Water signs are the intuitive dreamers of this world.*

CANCER
June 21 to July 22

RULING PLANET: MOON

BIRTHSTONE: MOONSTONE, PEARL

Cancer is a home-loving sign concerned with nurturing and the family. It is the sign of the mother. Moonstone reflects the intuitive quality of its ruler, the Moon. Moonstone's soothing qualities calm a tendency to emotional overreaction and touchiness. Moonstone affects the ebb and flow of the menstrual cycle and regulates that cycle. However, if worn, it may need to be removed at the highly charged full moon – Cancer's most sensitive time. Cancer has strong links with the past, and Pink Tourmaline aids acceptance of the past and enhances trust.

MOONSTONE

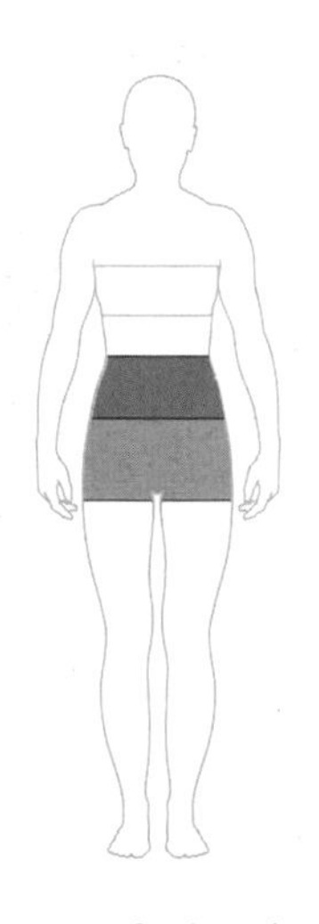

ABOVE *In medical astrology, Cancer rules the breasts and stomach and, through its ruler the Moon, the female reproductive system.*

AFFINITIES

Amber, Beryl, Calcite, Chalcedony, Chrysoprase, Emerald, Moonstone, Opal, Pink Tourmaline, Rhodonite, Ruby

SCORPIO
October 23 to November 21

RULING PLANETS: MARS, PLUTO
BIRTHSTONES: TOPAZ, MALACHITE

Magnetic Scorpio is an intense and enigmatic sign. Powerful emotions and strong sexual feelings flow beneath an outwardly calm surface. Known for its perspicacity, Scorpio is one of the signs where transformation occurs, a quality of its ruler Pluto. Malachite can promote that transformation. It resonates with loyalty and fidelity, Scorpio attributes. With strong links to the occult, Scorpio benefits from the protective energies of crystals such as Apache Tears. Garnet helps Scorpio to direct the spontaneous rise of sexually charged kundalini energy. Topaz helps Scorpio to overcome resentment and learn forgiveness.

TOPAZ

MALACHITE

AFFINITIES

Apache Tears, Aquamarine, Beryl, Boji Stone, Charoite, Dioptase, Emerald, Garnet, Green Tourmaline, Herkimer Diamond, Kunzite, Malachite, Moonstone, Obsidian, Rhodochrosite, Ruby, Topaz, Turquoise, Variscite

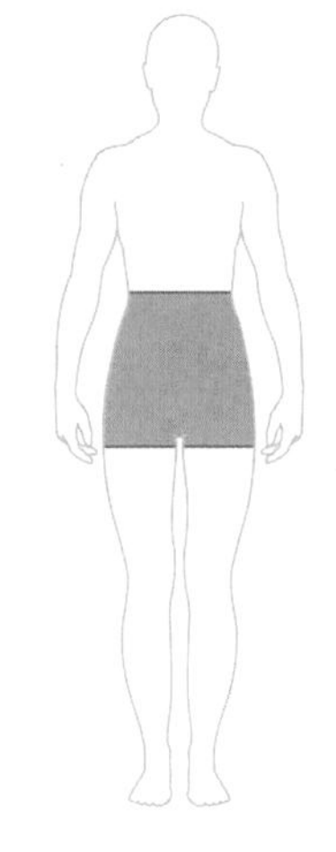

ABOVE *In medical astrology, Scorpio rules the reproductive system, the bowels and the bladder.*

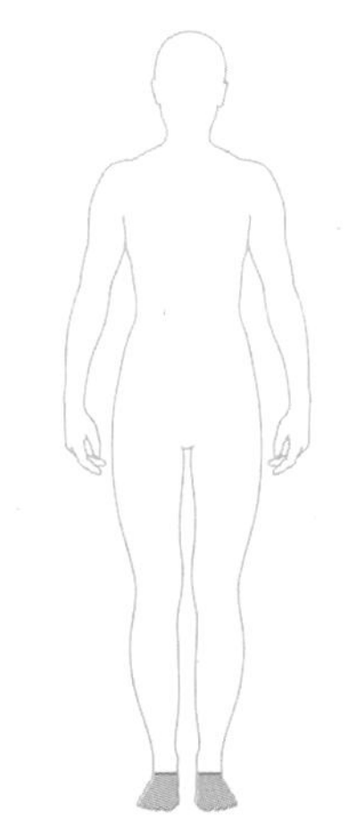

ABOVE *In medical astrology, Pisces rules the feet. Pisceans often need a grounding stone to keep their feet on earth!*

PISCES
February 19 to March 20

RULING PLANETS: JUPITER, NEPTUNE
BIRTHSTONES: MOONSTONE, AMETHYST

Sensitive and intuitive, Pisces people swim through life as the mood takes them, responding to hidden emotional currents. Ruler Neptune is imaginative and mystical but lacks boundaries, often making Pisces a "psychic sponge" that soaks up feelings and emotions from others. Amethyst aids Pisces to cleanse these energies and create workable boundaries. It also provides a link between everyday reality and the psychic reality of the spiritual world, a world that Pisces is in tune with. Labradorite, too, acts as a bridge between the worlds. Fluorite imparts discernment and detachment, useful qualities for Pisces. Moonstone heightens awareness of the effects of emotions and unconscious feelings.

AMETHYST

MOONSTONE

AFFINITIES

Amethyst, Aquamarine, Beryl, Bloodstone, Blue Lace Agate, Calcite, Chrysoprase, Fluorite, Labradorite, Moonstone, Turquoise

Air crystals

Air shapes ideas. This is the world of intellectual perception and logical thought. Air is the element of discernment and analysis, of comment and dissection. It is the mental element that thinks things through but also has flashes of intuition. The Air element is inventive, full of new ideas. Air signs need to express themselves verbally, for this is the element of communication. The Air signs are Gemini, Libra, and Aquarius.

GEMINI
May 21 to June 20

RULING PLANET: MERCURY

BIRTHSTONES: TOURMALINE, AGATE

Lively Gemini is ruled by Mercury. This talkative sign simply has to communicate. Multi-tasking Gemini has a natural adroitness that is reflected in its Agate birthstone. Agate's ability to stimulate precise, analytic attention to detail can help Gemini to overcome a tendency to pass lightly over inconvenient facts. Apophyllite attunes Gemini to truth. Restless Gemini can be prone to nervous tension, and Green Tourmaline provides an antidote to mental stress. Gemini can also benefit from calming Dendritic Agate to remain focused during confused or disorientating situations.

RIGHT *Talkative Air signs may need a crystal to counteract the emissions of that mobile phone*

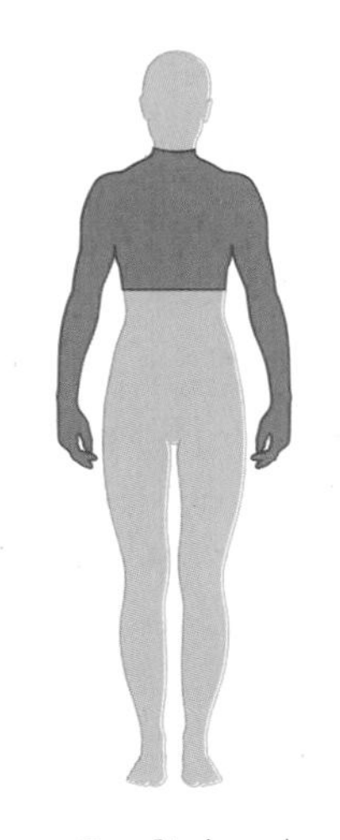

ABOVE *In medical astrology, Gemini is connected to the hands and arms, the lungs and respiratory system, and the nervous system.*

BLUE LACE AGATE

AFFINITIES

Agate, Apophyllite, Aquamarine, Calcite, Chrysocolla, Chrysoprase, Citrine, Dendritic Agate, Green Tourmaline, Sapphire, Serpentine, Tourmalinated and Rutilated Quartz, Tiger's Eye, Topaz, Variscite

GREEN TOURMALINE

LIBRA
September 23 to October 22

RULING PLANET: VENUS

BIRTHSTONES: SAPPHIRE, OPAL

The sign of marriage and partnership, Libra is happiest in relationships: not surprisingly, since amorous Venus rules this charming sign. Peace and balance are keynotes of Libra. This sign is concerned with beauty of all kinds, and a Sapphire birthstone promotes peace and harmony of mind. Libra particularly resonates with White Sapphire, a stone that supports Libran qualities of morality, justice, and freedom. Libra also has a strong affinity with Aquamarine, which softens the judgmental side of the sign. Aquamarine allows energy to flow in a structured, balanced way, aligning chakras and the physical and etheric bodies.

OPAL

SAPPHIRE

AFFINITIES

Apophyllite, Aquamarine, Aventurine, Bloodstone, Chiastolite, Chrysolite, Emerald, Green Tourmaline, Jade, Kunzite, Lapis Lazuli, Lepidolite, Mahogany Obsidian, Moonstone, Opal, Peridot, Sapphire, Topaz

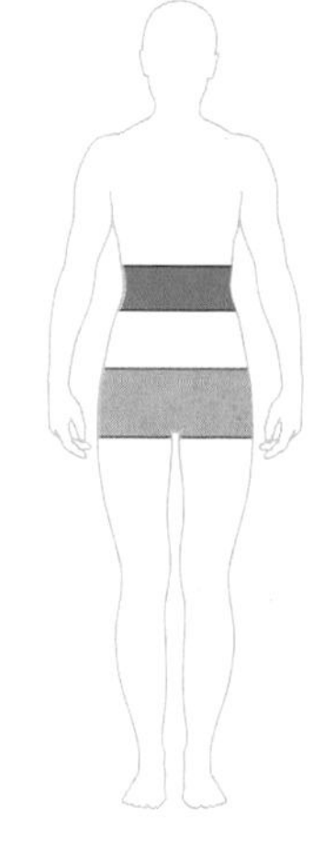

ABOVE *In medical astrology, Libra rules the kidneys and buttocks.*

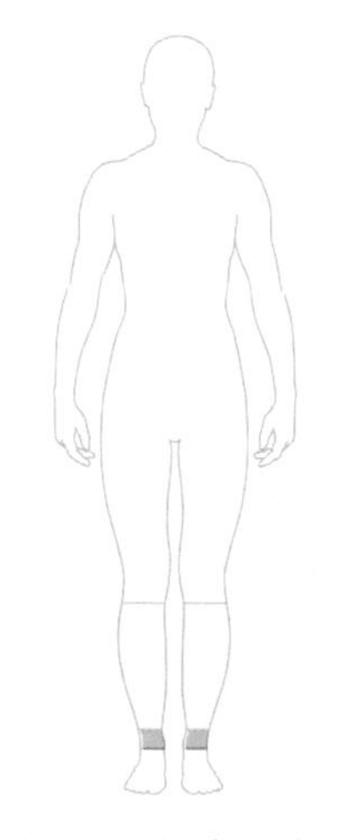

ABOVE *In medical astrology, Aquarius rules the calves and ankles.*

AQUARIUS
January 20 to February 18

RULING PLANETS: SATURN, URANUS

BIRTHSTONES: AQUAMARINE, AMETHYST

Attuned to revolutionary Uranus, Aquarius is the most way-out sign. There is something quirky and eccentric about this far-sighted, humanitarian sign. It is as though Aquarians not only see into the future but want to bring it about now. Amethyst aids in this goal, assisting in assimilation of new ideas. Since the sign is ruled by Uranus, planet of change, and Saturn, planet of consolidation, Aquarius can be pulled in two directions at once. Both Blue Celestite and Angelite aid in balancing and aligning direction and purpose. In addition Blue Celestite enhances telepathic communication and promotes brotherhood throughout the universe, so it helps Aquarius reach out beyond the confines of the known world.

AQUAMARINE

AMETHYST

AFFINITIES

Amber, Amethyst, Angelite, Aquamarine, Blue Celestite, Boji Stone, Chrysoprase, Fluorite, Labradorite, Magnetite, Moonstone

Fire crystals

Fire shapes inspiration. This is the impetuous element of spirit and creative life. The spontaneous Fire element is active and outgoing. It is innovative and initiatory. Fire signs get things moving. They act immediately with verve and confidence. A passionate element, this is the world of inspiration and creativity. The Fire signs are Aries, Leo, and Sagittarius.

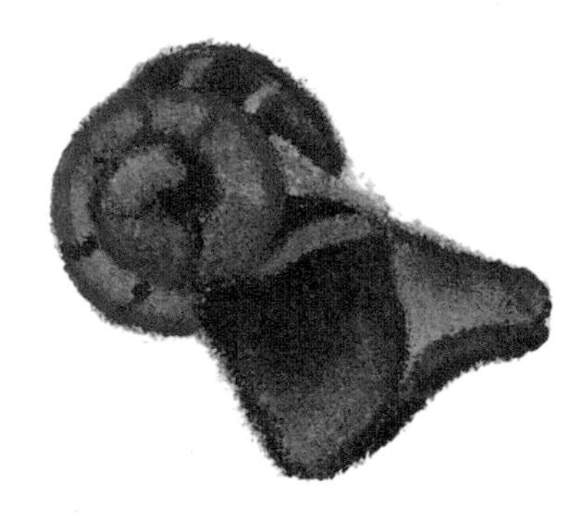

RIGHT *Fire signs are passionate and loving, confident and innovative.*

ARIES
March 21 to April 19

RULING PLANET: MARS

BIRTHSTONES: RUBY, DIAMOND

The first sign of the Zodiac, Aries is "me" oriented. Its keynotes are action and courage. This is a brash, pushy sign refracting its ruler, the warrior, Mars. Fearless Mars is assertive and aggressive, and so is Aries. The Ruby birthstone needs to be handled with care. It can bring anger to the surface to be burned away. It is an energizing stone and can overwhelm with passion, something Aries has in abundance. Used positively, Ruby can help impetuous Aries to focus goals and aspirations more wisely. Carnelian helps inflammatory Aries to avoid dangerous situations, while Amethyst soothes and calms the impatient Aries nature. Diamond helps Aries to be more thoughtful and considerate.

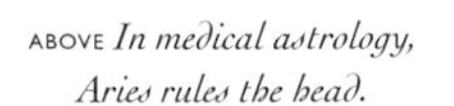

ABOVE *In medical astrology, Aries rules the head.*

AFFINITIES

Amethyst, Aquamarine, Aventurine, Bloodstone, Carnelian, Citrine, Diamond, Fire Agate, Garnet, Jadeite, Jasper, Kunzite, Magnetite, Pink Tourmaline, Ruby, Topaz

DIAMOND

RUBY

LEO

July 23 to August 22

RULING PLANET: SUN

BIRTHSTONES: CAT'S (OR TIGER'S) EYE, RUBY

Leo is the natural ruler of the Zodiac. The lion has a larger-than-life personality. Warmth and exuberance are Leonine keynotes. The Sun lights up life for open-hearted Leo – and everyone around them. This proud sign wants to be noticed and it is difficult to ignore a dramatic Leo, especially when bedecked with the gold jewelry and yellow crystals of its solar ruler. The Cat's (or Tiger's) Eye birthstone can help Leo find positive expression for an abundant fund of energy. Leo responds to all the yellow crystals, but Citrine, like Leo, is particularly brilliant and energizing. It aids Leo in harnessing creative energy. Ruby resonates with the generosity and enthusiasm of Leo and brings out inherent leadership qualities.

CAT'S EYE

AFFINITIES

Amber, Boji Stone, Carnelian, Cat's Eye, Chrysocolla, Citrine, Danburite, Emerald, Fire Agate, Garnet, Golden Beryl, Green and Pink Tourmaline, Kunzite, Muscovite, Onyx, Orange Calcite, Pyrolusite, Quartz, Rhodochrosite, Ruby, Topaz, Turquoise

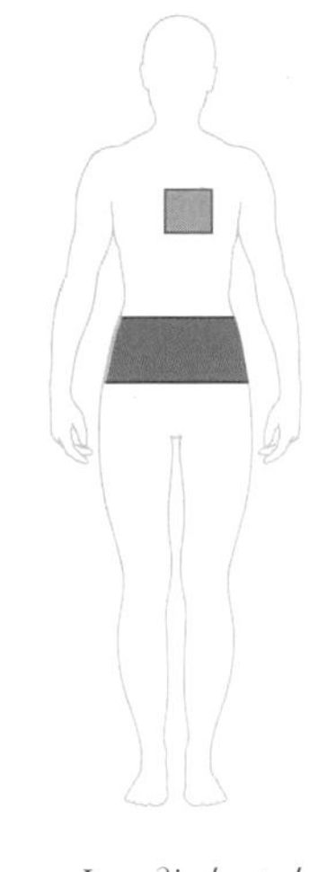

ABOVE *In medical astrology, Leo rules the heart and the lower back.*

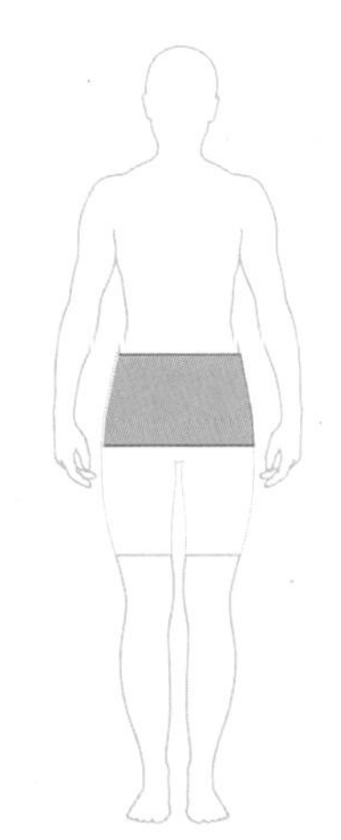

ABOVE *In medical astrology, Sagittarius rules the hips and thighs. Charoite is useful in helping to overcome problems in these areas.*

SAGITTARIUS

November 22 to December 21

RULING PLANET: JUPITER

BIRTHSTONES: TOPAZ, TURQUOISE

Sagittarius is engaged on an eternal quest for meaning, a quest that is supported by the golden rays of Topaz. Topaz aids vision, broadens perspective, and lights up the path ahead. Its philanthropic urges are in tune with Sagittarius's leader, beneficent Jupiter. Lapis Lazuli, the "stone of total awareness," helps Sagittarius to expand under Jupiter's tutelage while avoiding excess. Azurite, too, can help Sagittarius. This stone promotes the relaxation required for meditation. It focuses inner vision and shuts off the questioning mind, creating a space for Sagittarius to simply be. This fiery sign tends to go all out toward a goal. Azurite helps prevent burnout, while Turquoise helps this fiery sign achieve serenity and peace.

TOPAZ

TURQUOISE

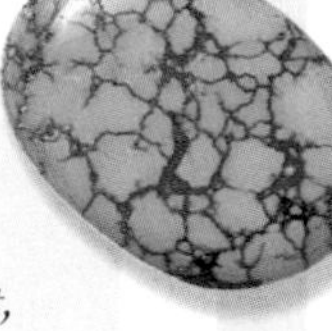

AFFINITIES

Amethyst, Azurite, Blue Lace Agate, Chalcedony, Charoite, Dioptase, Garnet, Labradorite, Lapis Lazuli, Malachite, Pink Tourmaline, Ruby, Smoky Quartz, Sodalite, Topaz, Turquoise

8 Crystal Wisdom

Crystals have had a long life upon the earth. They have witnessed evolution. Some have undergone many changes, others retain their pristine form. One or two, such as Moldavite and meteoric Magnetite, have come from outside our solar system. Some people believe that they bring a message from the stars.

With their ability to absorb energy, crystals hold knowledge and pass it on to those who are ready to hear. Crystals want to share their wisdom; it is part of their purpose in emerging from the earth. This is why they formed part of sacred practices for years.

Sleeping with a wisdom crystal under your pillow endows you with prophetic dreams. If you sit quietly and contemplate crystals, they whisper their secrets. They help you to know yourself fully. Crystals stimulate healing powers and help you know the past and the future. Tuning into crystals sharpens intuition.

The intuition is like an opening flower, an idea reflected in the petals of this delicate Amethyst.

Attuned to a fine vibration, crystals bring celestial knowledge to earth. Meditating with crystals expands your consciousness and facilitates contact with the angelic realms. Contemplation helps you to understand the properties of crystals more fully and develops your sensitivity.

As we enter the next millennium, many people believe that the earth will undergo great changes. New insights and meaning are found in crystals to aid this process. Crystals such as Quartz, known about since time began, are revealing new facets of themselves. "Record-keeper" crystals are emerging from the earth to tell their stories and share sacred knowledge. "Teacher" crystals share their knowledge with students who have been specially called.

Crystals for wisdom

Crystals have an inherent wisdom. They store knowledge and release information when quietly attuned to. Sitting with your crystals brings you wisdom, insight, and vision – especially when they are placed on the forehead: on your third eye. They can link you to the knowing hidden in the depths of your being, or they can take you far beyond the solar system to access undreamed-of wisdom from the stars.

ABOVE *Moldavite is believed by some people to have been sent from outside our solar system to bring a vision from the stars of what could be. It is certainly of extraterrestrial origin.*

STONES OF VISION

Peridot

Peridot is a pure gemstone of great clarity. It is a mix of yellow buffered with blue, creating light green. Peridot aids in understanding destiny and the purpose of existence. It puts you in touch with ultimate truth.

Moldavite

The only known gem-quality stone of extra-terrestrial origin, Moldavite reaches out beyond earth. It expands vision and the ability to attune to those wider vibrational frequencies. Moldavite takes you into endless possibilities.

BELOW *Crystals are washed out from the earth when their knowledge or properties are needed. Many crystals carry insight and offer visions.*

STONES OF ANCIENT WISDOM

Serpentine

SERPENTI

Serpentine is an old stone and links back into ancient wisdom, helping to understand the spiritual aspect of existence.

Snow Quartz

Snow Quartz is milky and opaque, often found as water-smoothed beach pebbles. Unlike Clear Quartz, it does not let you see into its depths. However, it does assist in seeing into your own depths. Snow Quartz aids in retrieving wisdom buried deep within yourself.

Herkimer Diamond

Herkimer Diamond is a particularly clear bright Quartz with rainbows in its depths. Another of the ancient stones, it has an in-built store of ecological knowledge and can facilitate recalling your own past lives.

Record-keeper Crystals

A long, thin Quartz crystal with one perfect-ly smooth side on which are engraved pyramid or hieroglyphs is a record-keeper. It has special knowledge to bring.

STONES OF INSIGHT

Azurite

A deep blue stone used by priestesses of Egypt to enhance their spiritual consciousness. Azurite is copper-based with a metallic sheen, the color of the desert night sky. Facilitates psychic abilities; enhances intuition; shows you your own innate resources.

Yellow Calcite

The waxy sheen of translucent Yellow Calcite lifts the spirit. It is connected with the light of the sun, traditionally linked with spiritual purity. Yellow Calcite aids meditation and focuses concentration during psychic work.

Emerald

Ranges from very light to very dark green. The transparent, light green stone helps with meditation, aligning physical, mental, and emotional bodies. Brings wisdom from the mental plane, which can be passed to others.

Fluorite

Fluorite usually crystallizes in a square or diamond form. It assists spiritual awakening, helping you to attain high levels of reality, as it anchors the energies into the everyday world. Colorless Fluorite opens the crown chakra, linking the physical to the spiritual realm. It facilitates recognition and release of anything not conducive to spirituality.

Chrysolite

Another stone linked to the sun, Chrysolite increases spiritual receptivity, heightens inspiration, and relieves negative mind states.

Sapphire

The best Sapphire for meditation is the deep, dark blue of the night sky. Its peace and serenity remove distractions and draw off negative energy. It clears spiritual confusion and puts you in touch with your essence.

Labradorite

The iridescent blue flashes in the depths of Labradorite elevate consciousness, bringing great insight without great searching. Its ability to protect the aura makes it invaluable during meditation.

Lapis Lazuli

Another of the sacred stones of Egypt, Lapis Lazuli is flecked with Iron Pyrite. Its extraordinarily serene deep blue is the color of spirituality. It is used to attain enlightenment but cannot go beyond the stage of spiritual evolution its wearer has achieved.

Turquoise

Turquoise unites the earth and sky, making it particularly valuable in spiritual work. It aids the search for who you really are.

BELOW *A few crystals have dropped in from outside our solar system to bring their wisdom to earth.*

BELOW *The goddess of wisdom and compassion, Kuan Yin, is made here from Turquoise, a stone that unites divine energies with those of earth.*

SAPPHIRE

LABRADORITE

LAPIS LAZULI

Crystal meditation

Meditation is a way of shutting off mind chatter, stilling the mind. It has many benefits, including alleviating stress and lowering blood pressure, but it also allows you to get to know your crystals and what they can do. In the stillness of meditation, the crystals will talk to you.

MEDITATING WITH YOUR CRYSTALS

Meditating with crystals is like opening a door to another world, especially if you choose a crystal that has fault lines and occlusions within it. You lose yourself within the crystal, and in the peace that follows, solutions and insights rise up into awareness. It is beneficial to meditate with each of your crystals in turn, taking a few days to tune into each one and come to know it fully.

ABOVE *Meditating through the spectrum of colors found in the crystal world progressively raises consciousness to the highest level.*

Some people like to have a "crystal day" when they tune into each of their crystals. Start with the red crystals to energize and awake, moving through the rainbow spectrum into orange, yellow, green, blue, purple, and violet. This brings you to the highest crystal vibration and you could well feel "blissed out." You may need to earth your energies again with Smoky Quartz, Boji Stone, or one of the black crystals. Earthing your energies after meditation is important, otherwise you feel "floaty and not quite here" – not a good space to be in when dealing with the everyday world.

HOW TO MEDITATE

Settle yourself comfortably with your crystal. Breathe gently, letting each out-breath be a little longer than the in. As you breathe out, let go of any stress or tension you feel. As you breathe in, let peace flow with it. Allow your breathing to settle into an easy rhythm. With softly focused eyes, look at your crystal.

RIGHT *A crystal that has internal planes is an excellent focus for meditation; it quickly brings about a visionary state.*

Notice its color, its shape, its weight if you are holding it. Allow yourself to wander within the crystal, exploring its inner planes.

When you are ready, close your eyes. Quietly contemplate the energies of the crystal and let it teach you about itself.

When you have finished, open your eyes. Place feet firmly on the floor and take a few moments before getting up and moving around. To ground, hold a Smoky Quartz.

ABOVE *Playing with crystals and allowing them to form a mandala helps you to focus your concentration and to absorb the crystal energies.*

MEDITATION-ENHANCING CRYSTALS

A CRYSTAL MANDALA

A mandala is a sacred form or pattern. Its repeating elements focus and calm the mind, bringing about a change of consciousness. A wheel with rim and spokes outlined in one color and spaces filled in with another is a mandala. If you find a sitting meditation difficult to settle to, then the moving meditation of a mandala may be more appropriate. You can create a mandala with tumbled stones, long or short crystals, or crystal clusters.

RIGHT *A large Apophyllite cluster is a wonderful aid to meditation as you journey through its translucent spheres.*

Contacting the angelic realms

Many people seek contact with the angelic realms. They reach out to a guardian angel for guidance and protection, or to angelic forces for inspiration and enlightenment. Crystals with a fine resonance and pure color facilitate this process. They can be put under the pillow at night or held in the hands during quiet moments of attunement. If the crystal has a "gateway," an opening leading into the crystal, it leads deeper into the angelic realm.

ABOVE *Placing a polished Selenite crystal under your pillow at night not only keeps you safe and aids restful sleep, but also attracts your guardian angel.*

CELESTITE

Celestite can enhance intuition and improve contact with the spirit realm, the angelic forces and guardian angels. Clairvoyant or clairaudient messages that are received through Celestite are clear and lucid. Celestite also has an inherent wisdom that promotes spiritual advancement and discernment.

This crystal can aid purposeful astral travel and improve dream recall, especially when those dreams are related to spirit or angelic guidance.

Celestite

Celestite is a beautiful sulfate mineral whose name means celestial. The most favorable color for connection with the angelic realms is the exquisite, pellucid light-blue. Its ethereal color immediately invokes the celestial realms and for this reason it is often known as Angelite. Angelite is a compressed form of Celestite and its banding often resembles angel's wings. However, Celestite color ranges through white, yellow, and orange to red and reddish-brown, all of which have powerful healing and balancing properties.

ANGELITE

SELENITE

Selenite

Selenite is formed from crystallized gypsum. Its outer appearance is translucent, like fine-ribbed ice, suggesting purity and higher consciousness. It is named after the moon and can have the color of pure white moonlight. When cut and polished, it is transparent, with delicate striated lines or half-hidden holographic images. Polished Selenite may also suggest an interior icescape. It makes an excellent focus to quiet the mind.

Selenite feels transcendent. It is pure and serene, radiating calmness and inner light. It brings great clarity of mind and sharpens spiritual contact, giving clear judgment. Meditating with Selenite brings you into a new dimension of consciousness. Selenite can also be used for past-life recall and for projecting into a possible future life. In psychic protection, Selenite helps to cut entity attachment.

Muscovite

Muscovite is a form of mica. Often pale and clear, Muscovite can range in color from pearl-like white, pink, or gray to yellow, green, or brown. It may also appear as violet or rose-red.

MUSCOVITE

Muscovite facilitates astral projection and permits contact with the spiritual realms. A visionary crystal, it can bring you into contact with both the angelic forces and a higher aspect of yourself. This is a reflective crystal, enabling you to look within. It throws light on your projections – the parts of yourself that you do not recognize and therefore see "out there" in someone else. With Muscovite, you learn to love the other in yourself. Muscovite can aid decision-making since it increases intuition and strengthens access to higher guidance.

ADDITIONAL CRYSTALS FOR ANGELIC CONTACT

✤ *Angelite (see opposite)* ✤ *Aquamarine* ✤ *Danburite* ✤ *Morganite (Pink Beryl)*

MORGANITE

DANBURITE

AQUAMARINE

LEFT *The gateway in this Selenite pillar leads directly into the angelic realms.*

Crystal Directory

There are many shapes and colors of crystal, some well known, others unfamiliar. One crystal can occur in a range of colors. You may have a crystal someone has given you but whose name you do not know. Or maybe you found one when you were out walking. Before you can gain maximum benefit from the crystal, you need to identify it. The pages that follow will help you to do this.

The directory is color-coded. Each crystal has one main entry, under its typical color (cross-references will lead you to the right place). Sometimes, however, the crystal's properties change according to the color it appears in. For example, the short entry Pink Carnelian, besides leading you to the main entry, Carnelian, under orange, will also give you the additional properties specific to pink.

To begin identification, look carefully at your crystal. Polishing or tumbling can affect its appearance. Some crystals have a translucent, crystalline structure; others are opaque or solid. One crystal may be smooth, another gritty in texture. All crystals have a basic color or combination of colors. Some are banded, others spotted. In the pages that follow, photographs help you recognize the crystal. Major sources of the crystal are listed so that if, for instance, you pick up a crystal when out walking, you will be able to eliminate certain crystals as you identify your find.

The list of properties under each crystal will help you to select appropriate crystals for yourself or to give as gifts to friends. You will also find a suggestion as to where to place or wear the crystal for maximum benefit.

Tumbling and polishing or faceting can change the appearance of a crystal, especially a gemstone.

BROWN CRYSTALS

FIRE AGATE

SOURCE USA, Czech Republic, India, Iceland, Morocco, Brazil
APPEARANCE Slightly iridescent
PROPERTIES Deep connection to earth. Calms and settles energy, especially before meditation. Aids introspection; brings inner problems and blocks up for examination, slowly and safely, dispelling fear. Reflects harm back to source so it may understand the effect. Strong sexual connection. Alleviates stomach and endocrine problems, treats circulatory disorders, central nervous system, and eye problems, and enhances night vision.
See Agate, page 100, for general information.
POSITION Wear anywhere, especially on the forehead.

AMBER

SOURCE Britain, Poland, Italy, Romania, Russia, Germany, Burma, Dominica
APPEARANCE Transparent, insects trapped inside
PROPERTIES Brings wisdom, balance, patience, and promotes altruism. Useful healer, draws disease from body, aids tissue revitalization. Heals nervous system, promotes self-healing. Elixir is antibiotic. Eases stress, neutralizes negative energy. Cleanses environment, body, mind, and spirit. Aids depression, memory, decision-making. Heals throat and throat chakra, kidneys, bladder.
POSITION Wear on the wrist and throat.

BOJI STONE™

SOURCE USA, Britain
APPEARANCE Metallic-looking, smooth or with square protrusions. The round form is considered to be female, the square form male
PROPERTIES Grounds back into reality after spiritual activity. Protective function. Aligns spiritual bodies and chakras. Clears blockages. Restores health.
POSITION Hold or carry.

HAWK'S EYE

SOURCE USA, South Africa, India, Mexico, Australia
APPEARANCE Banded, "hawklike eye"
PROPERTIES Form of Tiger's Eye. Earths energy. Stimulating and energizing. Attracts abundance. Aids circulatory system, bowels, and legs.
POSITION Hold or place on the appropriate point.

MAGNETITE (LODESTONE)

SOURCE USA, India, Mexico, Romania, Italy, Finland, Austria
APPEARANCE Dark and grainy (iron ore)
PROPERTIES Aids telepathy, visualization, and balanced perspective. Connects to nurturing aspect of earth. Can attract love. Beneficial for blood and circulatory system. Stimulates sluggish organs and sedates the overactive. Useful for sports injuries; relieves aches and pains.
POSITION Place on the back of the neck and the base of the spine, or on an aching joint.

BROWN CRYSTALS

SMOKY QUARTZ

SOURCE USA, Britain, Brazil
APPEARANCE Translucent, long, pointed crystals, usually darker at points
PROPERTIES Aids elimination and detoxification on all levels. Relieves fear; lifts depression, bringing calmness and positive thought. Useful for radiation-related illness or chemotherapy. Aids acceptance of body, cleans base chakra. Aids virility. Grounds spiritual energy, neutralizes negative energy. Alleviates nightmares, manifests dreams.
See also Quartz, page 100.
POSITION Wear anywhere, especially on the base chakra. Place under the pillow at night.

TIGER'S EYE

SOURCE USA, Mexico, India, Australia, South Africa
APPEARANCE Banded
PROPERTIES Protective stone, shows correct use of power. Aids accomplishing goals, recognizing inner resources. Brings clarity of intention. Anchors change into physical body. Integrates brain hemispheres, balances yin and yang energy, energizes body. Enhances psychic abilities.
POSITION Wear on the right arm.

TOPAZ

SOURCE USA, Mexico, Brazil, Sri Lanka, Nigeria, Namibia, Ireland
APPEARANCE Transparent, pointed crystals
PROPERTIES Directs energy. Soothes, heals, stimulates, recharges, and remotivates, raising energy. Promotes truth and forgiveness. Sheds light on path, taps into inner resources. Brings joy and generosity, manifestation of good health. Cleans aura.
POSITION Wear on the ring finger.

SARDONYX

SOURCE Brazil, India, Russia, Asia Minor
APPEARANCE Banded
PROPERTIES Associated with strength and protection. Increases stamina and vigor, aids self-control. May bring lasting happiness in marriage, attract friends, lift depression. Heals lungs and bones. **Brown Sardonyx** is particularly important in grounding energy.
POSITION Anywhere.

CHIASTOLITE (CROSS STONE)

SOURCE China
APPEARANCE Dark cross in center
PROPERTIES Dispels negativity. Transmutes conflict into harmony; aids change, problem-solving. Brings creativity, answers to mysteries. Signifies death and rebirth. Facilitates astral travel. Lessens fevers, repairs chromosome damage.
POSITION Place where appropriate or wear around the neck.

BROWN CRYSTALS

MAHOGANY OBSIDIAN

SOURCE Worldwide
APPEARANCE Deep brown and glassy
PROPERTIES Grounding and protecting, brings strength in times of need, vitalizes purpose. Eliminates energy blockages, relieves tension. Stimulates growth on all levels.
See also Obsidian, page 123.
POSITION Hold or place on the appropriate point.

MOSS AGATE

SOURCE USA, Australia, India
APPEARANCE Branching markings, like moss
PROPERTIES Cleanses circulatory and elimination systems of body. Eliminates depression caused by left and right brain imbalances. Helps intellectual people access intuitive feelings, loosening restrictions. Conversely, helps intuitive and creative people channel energy practically. Aids hypoglycemia. Elixir treats fungal infections.
See also Agate, page 100.
POSITION Hold or place on the appropriate point.

BROWN SELENITE

SOURCE Britain
APPEARANCE Brown, crystalline
PROPERTIES Helps to earth angelic energies.
See also Selenite, opposite.
POSITION Hold.

BROWN JADE

PROPERTIES Earthing stone, brings comfort and reliability. Aids in adjusting to new environment.
See also Jade, page 116.

BROWN SERPENTINE

See Serpentine, page 124.

BROWN TOURMALINE

PROPERTIES Clears aura, aligns etheric body, protecting it. Treats intestinal disorders. Also called Imperial Tourmaline.
See also Tourmaline, page 117.

BROWN MUSCOVITE

See Muscovite, page 105.

BROWN JASPER

PROPERTIES Encourages environmental awareness, facilitates deep meditation, regression, and centering. Gives night vision and aids astral travel.
POSITION Wear on the forehead.
See also Jasper, page 102.

WHITE MUSCOVITE

See Muscovite, page 105.

WHITE BERYL

See Beryl, page 114.

WHITE TOPAZ

See Topaz, page 97.

WHITE CELESTITE

See Celestite, page 110.

WHITE DANBURITE

See Danburite, page 105.

WHITE CHALCEDONY

See Chalcedony, page 125.

WHITE CALCITE

See Calcite, page 106.

WHITE LEPIDOLITE

See Lepidolite, page 121.

WHITE CRYSTALS

APOPHYLLITE

SOURCE Britain, Australia, India, Brazil, Czech Republic, Italy
APPEARANCE Cubic crystals
PROPERTIES Creates conscious connection between physical and spiritual realm, facilitates conscious astral travel. Promotes introspection and correction of imbalances. Aids seeing future, stimulates intuitive vision.
POSITION Hold or place on the third eye. For channeling, place on the third eye.

SELENITE

SOURCE USA, Russia, Austria, Greece, Poland, Germany, France
APPEARANCE Translucent with fine ribbing
PROPERTIES Brings clarity of mind, accesses angelic consciousness. Reaches other lives. Assists judgment and insight, showing inner workings. Aligns spinal column, promotes flexibility.
POSITION Hold.

OPAL

SOURCE Australia, Mexico, Peru
APPEARANCE Fiery, shifting colors
PROPERTIES Enhances cosmic consciousness, originality, and dynamic creativity. Fosters love and passion. Induces visions. Brings loyalty, faithfulness, spontaneity. Amplifies traits. Emotionally responsive, encourages emotional stability. Treats Parkinson's Disease. Strengthens memory. Treats infections and fevers, purifies blood and kidneys. Regulates insulin. Eases childbirth. **Fire Opal** is good for business.
POSITION Wear on the little finger.

MOONSTONE

SOURCE India, Australia
APPEARANCE Milky, translucent
PROPERTIES "Stone of new beginnings," connected to the moon and intuition. Soothes and calms emotional lability and overreactions. Reflective, makes conscious the unconscious, aids intuition and empathy. Feminine stone associated with psychic abilities and cleansing. Powerful effect upon female reproductive cycle, balances fluid imbalances, attunes to biorhythmic clock. Aids digestive system, assimilating nutrients and eliminating toxins. Alleviates degenerative conditions of skin, hair, eyes, and fleshy organs. Used for Premenstrual Syndrome, conception, pregnancy, and childbirth.
POSITION Wear on the finger or place on the appropriate body part.

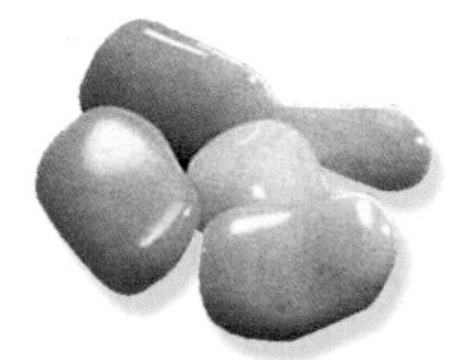

SNOW QUARTZ

SOURCE Worldwide
APPEARANCE Firmly compacted, opaque, milky; often found as water-worn pebbles
PROPERTIES Supports during the learning of lessons; especially connected with letting go of responsibilities and limitations that are no longer relevant. Enhances tact and cooperation in all situations. Links to deep inner wisdom that has been previously denied in both oneself and the society in which one lives.
See also Quartz, page 100.
POSITION Use anywhere.

CLEAR CRYSTALS

QUARTZ

SOURCE Worldwide
APPEARANCE Long, pointed crystals, some transparent, some milky or striated
PROPERTIES One of the most powerful healing stones. An energy amplifier; doubles auric field, storing, releasing, and regulating energy on physical and mental dimensions. Generates electromagnetic energy and dispels static electricity. Cleans and enhances subtle bodies. A deep soul cleanser. Dissolves karmic seeds. Enhances psychic awareness and amplifies psychic abilities. Connects physical dimension with mind, stimulates immune system. Soothes burns. Said to reduce fuel consumption in cars.
POSITION Use anywhere. Place on neck for thyroid; heart for thymus; solar plexus for energy; ears to balance and align.

TOURMALINATED QUARTZ

APPEARANCE Long, thick "threads" in clear crystal
PROPERTIES Dissolves crystallized patterns, harmonizing disparate and opposite elements. Brings harmony to polarities. A problem-solver.
See Quartz, above, for general information.
POSITION Use anywhere.

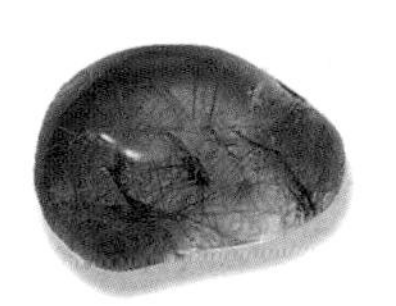

RUTILATED QUARTZ

APPEARANCE Long thin "threads" in clear crystal
PROPERTIES Gets to root of problems. Aids astral travel, absorbs mercury poisoning. Heightens energetic impulse of Quartz. Focuses non-physical atunement.
See Quartz, above, for general information.
POSITION Use anywhere.

DIAMOND

SOURCE South Africa, Australia, Brazil, India, Russia, USA
APPEARANCE Bright, transparent gemstone
PROPERTIES Symbol of purity; bonds relationships, bringing love and clarity. Amplifies energy. Qualities include fearlessness, invincibility, and fortitude. Aids glaucoma.
POSITION Wear on the finger or place on the temple.

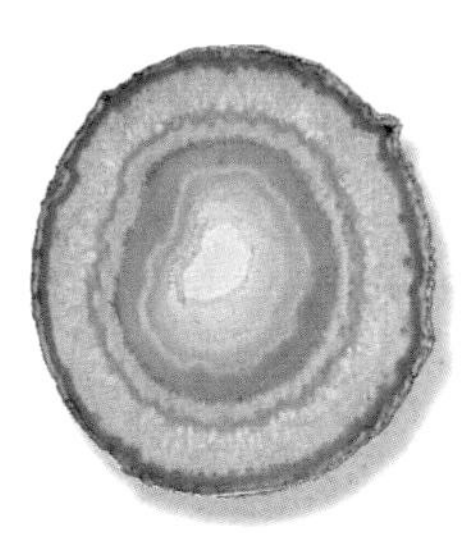

AGATE

SOURCE USA, India, Morocco, Czech Republic, Brazil, Africa
APPEARANCE Waxy and soft, often banded; often translucent (may be artificially colored)
PROPERTIES A grounding stone. Improves perception and analytical qualities. Balances yin and yang, and harmonizes physical, emotional, mental, and etheric qualities. Soothing and calming. Raises consciousness and builds self-confidence. Stabilizes aura, eliminates negativity. Overcomes bitterness of the heart and inner anger. Fosters love, truthfulness, and courage. Different types of agate have additional properties.
POSITION Hold or place on the appropriate point.

CLEAR CRYSTALS

DENDRITIC AGATE

SOURCE USA, Czech Republic, India, Iceland, Morocco, Brazil
APPEARANCE Fernlike markings
PROPERTIES "Stone of abundance." Heals connection with the earth; aids remaining centered in discordant situations. Treats skeletal disorders and nervous system, reverses capillary degeneration. Provides pain relief.
See also Agate, page 100.
POSITION Hold or place on the appropriate point.

HERKIMER DIAMOND

SOURCE USA, Mexico, Spain
APPEARANCE Clear, oily, inner rainbows
PROPERTIES Energizing and enlivening, promotes creativity. "Stone of attunement." Stimulates psychic abilities and dream recall. Detoxifier on all levels; protects against radioactivity, corrects DNA and metabolic imbalances, eliminates stress. Promotes past-life recall.
POSITION Wear in pendants or earrings, or place at the base of the spine or on the solar plexus.

CLEAR TOURMALINE

PROPERTIES Aligns meridians of physical and etheric bodies, opens crown chakra. Synthesizes all other colors.
See Tourmaline, page 117, for general information.
POSITION Hold or place on the appropriate point.

CLEAR FLUORITE

PROPERTIES Stimulates crown chakra, energizes aura, harmonizes intellect and spirit.
POSITION Wear on the earlobes. Place on a computer.
See also Fluorite, page 106.

CLEAR SARDONYX

PROPERTIES A purification stone.
POSITION Anywhere.
See also Sardonyx, page 97.

CLEAR DANBURITE

See Danburite, page 105.

RED CRYSTALS

RED JADE

PROPERTIES
Passionate, associated with love. Accesses anger, releases tension.

See Jade, page 116, for general information.
POSITION Hold or place on the appropriate point.

RED CALCITE

PROPERTIES
Increases energy, uplifts emotions, aids willpower. Opens heart chakra.

See Calcite, page 106, for general information.
POSITION Hold or place on the appropriate point.

GARNET

SOURCE Worldwide
APPEARANCE Transparent or translucent
PROPERTIES Energizes and revitalizes, balancing energy, bringing serenity. Activates other crystals, clears negative chakra energy. Inspires love, stimulates controlled rise of kundalini energy. Treats spinal and cellular disorders, blood, heart, and lungs; regenerates DNA. Aids assimilation of minerals and vitamins.
Red Garnet represents love. Attuned to heart energy, revitalizes feelings and enhances sexuality. Controls anger, especially toward oneself.
POSITION Wear on the earlobes, finger, or over the heart. For past-life recall place on the third eye.

JASPER

SOURCE Worldwide
APPEARANCE Patterned or banded
PROPERTIES Supreme nurturer. Brings tranquillity and wholeness, protection and grounding. Aids quick-thinking and organizational abilities, ability to see projects through. Aligns chakras, facilitates shamanic journeys and dream recall. Prolongs sexual pleasure.
Red Jasper grounds energy, rectifies unjust situations. Aids dream recall. A "stone of health." Strengthens circulatory system. Connected to base chakra, aids rebirthing.
POSITION Place Jasper on the forehead. Place Red Jasper on the base chakra.

RUBY

SOURCE India, Madagascar, Russia, Sri Lanka, Cambodia, Kenya
APPEARANCE Bright, transparent gemstone or cloudy crystal.
PROPERTIES Energizes and balances but may over-stimulate, bringing up anger or negative energy for transmutation. A leadership stone; stimulates the heart chakra and encourages "following your bliss." Shields against psychic attack, promotes positive dreams. Aids retaining wealth and passion. Detoxifies body and blood; treats fevers, restricted blood flow. Aids survival issues.
POSITION Wear over heart or on finger or ankle.

RED CRYSTALS

RED SARDONYX

PROPERTIES A stimulating stone.

See Sardonyx, page 97, for general information.

POSITION Wear or place anywhere.

CUPRITE

SOURCE USA, Britain, Germany, France, Namibia, Peru

APPEARANCE Small crystalline masses

PROPERTIES A philosophical stone, teaches how to be helpful to others. Overcomes difficulties in dealing with father or authoritarian figures. Aids past-life investigation. Stimulates base chakra, grounds energy, revitalizes physical energy. Aids heart and blood, muscle tissue, and skeletal system; overcomes metabolic imbalances. Treats water retention, bladder, and kidney malfunction; helps vertigo and altitude sickness.

POSITION Hold or place on the appropriate point.

AVENTURINE

SOURCE Italy, Brazil, China, India, Russia, Tibet, Nepal

APPEARANCE Opaque, microcrystalline

PROPERTIES Enhances creativity, brings prosperity. Diffuses negativity, balances male–female energy. Helps see alternatives and possibilities. Promotes growth during first seven years of life. Benefits thymus gland and nervous system. Balances blood pressure. Elixir aids skin problems.

POSITION Hold or place on the appropriate point.

RED CELESTITE

See Celestite, page 110.

RED FLUORITE

See Fluorite, page 106.

RED TOURMALINE

See Tourmaline, page 117.

RED LEPIDOLITE

See Lepidolite, page 121.

RED CARNELIAN

PROPERTIES Stimulates a weak voice.

See also Carnelian, page 109.

PINK CRYSTALS

KUNZITE

SOURCE USA, Madagascar, Brazil, Burma, Afghanistan
APPEARANCE Transparent or translucent, striated crystal
PROPERTIES Spiritual stone, produces loving thoughts and communication, radiates peace, and connects to universal love. Dispels negativity. Induces deep, centered meditative state, enhances creativity. Removes obstacles to path, aids self-expression, removing emotional debris and allowing free expression of feelings. Helps adjust to pressure of life, can heal psychiatric disorders. Activates heart chakra, aligns with throat and third eye. Shields from unwanted energies, dispels attached entities and mental influences from aura. Strengthens circulatory system and heart muscle.
POSITION Hold, or place on chakra.

ROSE QUARTZ

SOURCE South Africa, Russia, USA, Brazil, Japan
APPEARANCE Usually translucent but may be transparent
PROPERTIES Attunes to energy of unconditional love and heals heart. Calms and reassures, bringing about deep inner healing. Releases blockages from unexpressed emotions, opens heart chakra, soothes wounded heart and internalized pain. Helps those who have never received love to access heart. Aids positive affirmations for self-trust and self-worth. Teaches how to love oneself, encouraging self-forgiveness and acceptance. Aids physical heart and circulatory system. Said to increase fertility. Soothes burns and blistering.
POSITION Wear over the heart.

PINK AGATE

PROPERTIES Promotes love between parent and child.
See Agate, page 100, for general information.
POSITION Over the heart.

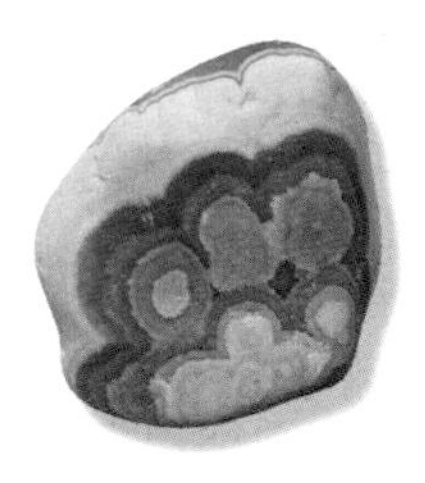

RHODOCHROSITE

SOURCE USA, Russia
APPEARANCE Banded
PROPERTIES A "stone of selfless love and compassion," improves feelings of self-worth, integrates spiritual and material energies. Expands consciousness, clears solar plexus chakra, enhances dream states. Alleviates irrational fear and paranoia. Teaches heart to assimilate painful feelings without shutting down, removes denial. Attracts soulmate. An irritant filter, aids asthma and respiratory problems. Purifies circulatory system and kidneys, restores poor eyesight. Elixir relieves infections, balances thyroid.
POSITION Wear on the wrist.

RHODONITE

SOURCE Spain, Russia, Sweden, Germany
APPEARANCE Mottled, often flecked with black
PROPERTIES Grounds, balances yin–yang, achieves highest potential. Shows both sides of an issue. Stimulates, clears, activates heart chakra, balancing physical and mental energies. Nurtures love. Useful for trauma. Aids confidence; alleviates confusion. Beneficially affects bone growth and hearing organs, fine-tuning auditory vibrations. Treats emphysema, inflammation of joints, arthritis. **Gem Rhodonite** activates pineal gland, brings intuitive guidance.
POSITION Hold or place as appropriate.

PINK CRYSTALS

PINK TOURMALINE

PROPERTIES Aphrodisiac. Brings love in material realm. Provides assurance it is safe to love, inspiring trust in love, confirms acceptable to love oneself. Helps share physical pleasure. Disperses emotional pain and old destructive feelings through heart chakra; synthesizes love and spirituality. Promotes peace. Balances a dysfunctional endocrine system.

See Tourmaline, page 117, for general information.

POSITION Place on the heart.

DANBURITE

SOURCE USA, Czech Republic, Russia, Switzerland, Japan, Mexico

APPEARANCE Transparent

PROPERTIES Activates intellect. Promotes ease in any situation; changes recalcitrant attitudes, bringing patience and peace of mind. Treats liver and gall bladder. Detoxifies, adds weight to body. Aids muscular and motor function.

Pink Danburite activates the heart chakra.

POSITION Place on the heart.

PINK CARNELIAN

PROPERTIES Promotes love between parent and child.

See Carnelian, page 109, for general information.

POSITION Wear as a pendant or belt buckle.

MORGANITE (PINK BERYL)

SOURCE USA, Brazil, Russia

APPEARANCE Crystalline, may be striated or pointed

PROPERTIES Attracts love and maintains it. Energizes loving thoughts and actions. Activates and cleanses the heart chakra. Brings calm in stressed life, benefiting nervous system. Oxygenates cells, reorganizing them; treats tuberculosis, asthma, and emphysema, and clears lungs.

POSITION Place on the heart.

MUSCOVITE

SOURCE Switzerland, Russia, Austria, Czech Republic

APPEARANCE Crystalline, pearl-like mica

PROPERTIES Strong angelic connection; stimulates heart chakra; facilitates astral travel. Disperses insecurity, self-doubt; aids intuition and problem-solving. Controls blood sugar, regulates kidneys.

POSITION Carry or hold.

YELLOW CRYSTALS

CAT'S EYE

SOURCE USA, South Africa, India, Mexico, Australia

APPEARANCE Banded or "eyelike," a form of Chrysoberyl

PROPERTIES Brings confidence, happiness, serenity, and good luck. Grounding stone, stimulates intuition, dispels negative energy from aura and protects it. Treats eye disorders, aids night vision. Relieves nervous headache and facial pain.

POSITION Wear on the right arm.

GOLDEN BERYL

PROPERTIES Promotes purity of being. Teaches initiative and independence, will to succeed, supporting potential. Opens crown and solar plexus chakras. Useful for ritual magic.

See Beryl, page 114, for general information.

POSITION Hold or place on the appropriate point.

CALCITE

SOURCE USA, Britain, Belgium, Czech Republic, Peru, Iceland, Romania

APPEARANCE Translucent and waxy

PROPERTIES Energy amplifier. Facilitates psychic abilities, channeling, astral projection, and higher consciousness. Connects intellect and emotions. Alleviates emotional stress, brings peace and serenity. Memory aid, brings insight. Cleanses organs and bones, strengthens skeleton and joints. **Golden or Yellow Calcite** enhances meditation, aids astral projection, focuses concentration. Brings in spiritual light and knowledge.

POSITION Hold or place on the appropriate point.

FLUORITE

SOURCE USA, Britain, Australia, Norway, China, Peru, Mexico

APPEARANCE Transparent, cubic crystals

PROPERTIES Protective stone, especially on psychic level. Protects against computer and electromagnetic stress. Aids physical and mental coordination, counteracting mental disorder, increasing concentration. Grounds and integrates spiritual energies. Promotes unbiased impartiality, heightens intuitive powers. Rekindles sexual libido. Benefits teeth, cells, and bones; repairs DNA damage. Aids arthritis and rheumatism. **Yellow Fluorite** enhances creativity, stabilizes group energy. Releases toxins.

POSITION Hold or place on the appropriate point.

YELLOW JADE

PROPERTIES Energetic and quietly stimulating, bringing joy. Teaches interconnectedness of all beings.

See Jade, page 116, for general information.

POSITION Hold or place on the appropriate point.

YELLOW CRYSTALS

YELLOW JASPER

PROPERTIES Protects during spiritual work and physical travel. Channels positive energy, energizes endocrine system.

See Jasper, page 102, for general information.

POSITION Forehead, chest, throat, and wrist.

CITRINE (CAIRNGORM)

SOURCE Britain, USA, Brazil, France, Madagascar, Russia

APPEARANCE Transparent crystal, may be cloudy

PROPERTIES "Stone of prosperity," attracts wealth and success. Brings happiness and generosity. Energizes and invigorates, increasing motivation and physical energy, activating creativity. Dissipates negative energy, promotes inner calm. Balances yin and yang, opens navel and solar plexus chakras, stimulates crown chakra. Cleanses aura and aligns etheric body with physical. Treats digestive problems, thyroid imbalance, and circulation of blood.

POSITION Wear on the fingers or at the throat.

YELLOW SAPPHIRE

PROPERTIES Particularly associated with the Hindu god of prosperity, Ganesh; attracts wealth. Stimulates intellect to give overall focus. Elixir removes toxins from body.

See Sapphire, page 119, for general information.

POSITION Touching the body or the finger.

GOLDEN YELLOW TOPAZ

PROPERTIES Recharges and strengthens faith and optimism. Attracts helpful people.

See Topaz, page 97, for general information.

POSITION Wear on the ring finger.

YELLOW CRYSTALS

YELLOW TOURMALINE

PROPERTIES Stimulates solar plexus, enhancing personal power. Opens up spiritual pathway.

See Tourmaline, page 117, for general information.

POSITION Hold or place on the appropriate point.

IRON PYRITE AND CUBIC PYRITE

SOURCE Britain, North America, Chile, Peru

APPEARANCE Metallic, may be cubic ("Fool's Gold")

PROPERTIES **Iron Pyrite** facilitates tapping into and unfolding unique abilities and potential. **Cubic Pyrite** expands and structures mental capabilities, balancing instinct and intuition, creativity with analysis. Useful when planning large business concepts; teaches how to see behind the façade to what is really there. Excellent energy shield, blocking out negative energy and pollutants, including infections. Treats bones and cell formation, repairs DNA damage, aids sleep disturbed due to gastric upset. Strengthens digestive tract, lessens ingested toxins. Benefits circulatory and respiratory systems, boosting transfer of oxygen from lungs to bloodstream.

POSITION Wear at the throat or place under the pillow.

YELLOW CELESTITE

See Celestite, page 110.

YELLOW KUNZITE

See Kunzite, page 104.

YELLOW MUSCOVITE

See Muscovite, page 105.

AMBER

See page 96.

YELLOW DANBURITE

See Danburite, page 105.

YELLOW GARNET

See Garnet, page 102.

YELLOW RHODONITE

See Rhodonite, page 104.

LEMON CHRYSOPRASE

See Chrysoprase, page 115.

ORANGE CRYSTALS

CARNELIAN

SOURCE Britain, India, Czech Republic, Peru, Iceland, Romania

APPEARANCE Translucent pebble, often water-polished

PROPERTIES Grounds and anchors into present surroundings. Removes fear of death, bringing acceptance of cycle of life. Improves analytic abilities and clarifies perception; motivates success in business; aids positive life choices; dispels apathy. Clears extraneous thoughts in meditation, heals etheric body, protects against rage and resentment. Capacity to cleanse other stones. Influences female reproductive organs, increasing fertility. Heals lower back problems, rheumatism, arthritis, depression. Good for dramatic pursuits; activates lower chakras.

Orange Carnelian warms and energizes.

POSITION Wear as a pendant or belt buckle.

ORANGE CALCITE

PROPERTIES Dissolves problems and maximizes potential.

See Calcite, page 106, for general information.

POSITION Hold or place on appropriate point.

ORANGE JADE

See Jade, page 116.

ORANGE CELESTITE

See Celestite, page 110.

ORANGE RHODOCHROSITE

See Rhodochrosite, page 104.

FIRE AGATE

See page 96.

PALE BLUE CRYSTALS

CELESTITE

SOURCE Britain, Egypt, Mexico, Peru, Poland, Libya
APPEARANCE Transparent, pyramidal crystals or planes
PROPERTIES Contacts spiritual and angelic realms; encourages spiritual development, enlightenment, and pureness of heart. Aids clairvoyant communication, dream recall, and astral travel. Disperses worries, attracts good fortune. Synthesizes instinct and intellect, aiding analysis of complex ideas. Brings balance and alignment. Treats disorders of eyes and ears, balances mind. Eliminates toxins, brings cellular order.
POSITION Hold or place on the appropriate point.

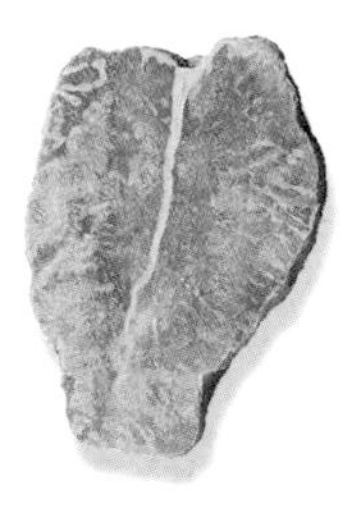

ANGELITE

SOURCE Britain, Egypt, Mexico, Peru, Poland, Libya
APPEARANCE Opaque and often veined
PROPERTIES Angelite is Celestite that has been compressed over millions of years. It shares many of the properties of Celestite. It also transmutes pain and disorder into wholeness and healing. Useful for angelic contact.
POSITION Hold or place on the appropriate point.

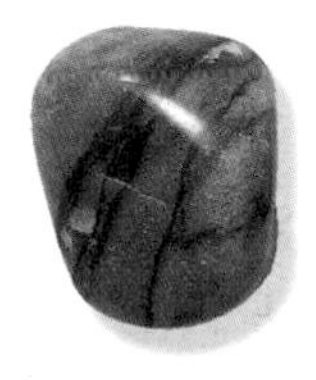

BLUE QUARTZ

PROPERTIES Assists in reaching out to others, assuaging fear. Blue Quartz which has been artificially coloured is known as Aqua Aura (see picture, pages 3 and 63) and, as the name suggests, it resonates with and cleanses the aura.
See Quartz, page 100, for general information.
POSITION Wear on the finger.

AQUAMARINE

SOURCE USA, Mexico, Russia, Brazil, India, Ireland, Zimbabwe, Afghanistan
APPEARANCE Clear crystal
PROPERTIES Useful healing stone, reduces stress, quietens the mind, sharpens intellect, clears blocked communication and throat chakra. Filters information reaching brain, aids interpretation of emotional state. Reduces fear, increases sensitivity and creativity, enhances intuition and spiritual awareness. Shields aura, aligns chakras. Brings tolerance to judgmental people, encourages responsibility for actions. Treats swollen glands and sore throats; strengthens cleansing organs, liver, spleen, and kidneys. Aids teeth, jaw, eyes, throat, and stomach.
POSITION Hold or place on the appropriate point.

BLUE LACE AGATE

PROPERTIES Calms, cools, lifts thoughts and takes spiritual inspiration to high vibration. Works on throat, heart, third eye, and crown chakras to bring about attunement. Treats arthritis and bone deformity, strengthens skeletal system, heals fractures. Aids blockages of nervous system, capillaries, pancreas. Elixir aids brain fluid imbalances and hydrocephalus.
See Agate, page 100, for general information.
POSITION Hold or place on the appropriate point.

PALE BLUE CRYSTALS

BLUE AVENTURINE

PROPERTIES A mental healer.

See Aventurine, page 103, for general information.

POSITION Anywhere.

BLUE FLUORITE

PROPERTIES Enhances orderly thought and clear communication, calms energy, and treats eye problems.

See Fluorite, page 106, for general information.

POSITION Hold or place on the appropriate point.

BLUE TOURMALINE

PROPERTIES Activates throat chakra and third eye. Aids psychic awareness, promotes visions. Opens way for service to others. Benefits pulmonary and immune systems, and brain.

See Tourmaline, page 117, for general information.

POSITION Hold or place on the appropriate point.

BLUE CALCITE
See Calcite, page 106.

BLUE TURQUOISE
See Turquoise, page 113.

BLUE JADE
See Jade, page 116.

BLUE OBSIDIAN
See Obsidian, page 123.

BLUE TOURMALINE
See Tourmaline, page 117.

MOSS AGATE
See page 98.

LIGHT BLUE SAPPHIRE
See Sapphire, page 119.

BLUE-GREEN CRYSTALS

CHRYSOCOLLA

SOURCE USA, Britain, Mexico, Chile, Peru, Zaire
APPEARANCE Opaque
PROPERTIES Tranquil and sustaining, aids meditation and communication. Assists speaking truth, gives personal confidence and sensitivity. Calms and re-energizes chakras. At heart chakra, heals heartache and increases capacity to love. At throat, improves communication. Aids keeping silent when appropriate. Treats arthritis and bone disease, muscle spasm, digestive tract and ulcers, blood disorders, and lung problems. Regenerates pancreas, regulates insulin and blood sugar balance.
POSITION Hold or place on the appropriate point.

BLUE-GREEN JADE

PROPERTIES Symbolizes peace and reflection, brings inner serenity. Aids people overwhelmed by situations beyond their control.
See Jade, page 116, for general information.
POSITION Hold or place on the appropriate point.

LABRADORITE

SOURCE Italy, Greenland, Finland, Russia
APPEARANCE Dull until catches light, then iridescent blue flashes
PROPERTIES Protective stone, deflecting unwanted energies from aura and preventing energy leakage. Aligns physical and etheric bodies, raises consciousness, grounds spiritual energies into body. Synthesizes intellectual thought with intuitive wisdom. Accesses spiritual purpose. Treats disorders of eyes and brain, relieves stress, regulates the metabolism.
POSITION Over the higher heart chakra.

BLUE-GREEN AGATE

APPEARANCE Artificially colored
PROPERTIES As with other Agates, Blue-Green Agate may be artificially colored. This does not give it special qualities but nor does it affect the general properties it has as an Agate. Artificially *made* Blue-Green Agate, however, is a glass and has no properties at all.
See Agate, page 100, for general information.

BLUE-GREEN CRYSTALS

TURQUOISE

SOURCE USA, Egypt, Mexico, China, Iran, Peru, Poland, Russia, France, Tibet
APPEARANCE Opaque, often veined
PROPERTIES Healing and protective, has elevating effect, promoting spiritual attunement. Purifying effect dispels negative energy. Unites energies of earth and sky, provides healing for spirit. Empathetic and balancing, uniting male and female. A promoter of self-realization, excellent for creative problem-solving. Strengthens the meridians of body and subtle energy fields; enhances physical and psychic immune systems. Alleviates pollution on all levels. Heals whole body, especially eyes.
POSITION Anywhere.

DIOPTASE

SOURCE Iran, Russia, Namibia, Zaire, Chile, Peru
APPEARANCE Transparent crystal, very deep green-blue
PROPERTIES Brings all chakras to higher level of functioning. Particularly useful in opening higher heart chakra and attaining spiritual attunement. Promotes living in present moment; activates past-life memory. Powerful physical and mental healing stone; regulates cell disorder; activates T-cells; relieves Ménière's disease; eases high blood pressure; alleviates pain including migraine. Said to prevent heart attacks and heal heart conditions.
POSITION At the higher heart chakra.

RAINBOW OBSIDIAN

PROPERTIES Enhances love, opens the spiritual side of one's nature, and is used for crystal gazing for relationships.
See Obsidian, page 123, for general information.
POSITION Hold or place on the appropriate point.

AQUA AURA

APPEARANCE Artificially coloured Quartz (illustrated on pages 3 and 63). Clear or cloudy, depending on Quartz base. May be speckled with gold (which does not rub off) and iridescent
PROPERTIES Cleanses aura, activates chakras, especially throat. Releases negativity from the subtle bodies. Creates space for something new. Enhances healing qualities of other crystals. Useful focus for meditation; deepens spiritual attunement and communication.
See Quartz, page 100, for general information.
POSITION Place as appropriate, or hold.

BLUE-GREEN AQUAMARINE

See Aquamarine, page 110.

BLUE-GREEN TOPAZ

See Topaz, page 97.

BLUE-GREEN OBSIDIAN

See Obsidian, page 123.

GREEN CRYSTALS

MOLDAVITE

SOURCE Czech Republic
APPEARANCE Small, transparent crystals
PROPERTIES Meteoritic, of extraterrestrial origin. Aids communication with higher self and with extraterrestrials. Used on the crown chakra, opens to higher spiritual energies; on throat, communicates interplanetary messages.
POSITION Place on forehead or crown.

GREEN FLUORITE

PROPERTIES Grounds excess energy, dissipates emotional trauma and clears infections. Aids stomach disorders and intestines.

See Fluorite, page 106, for general information.
POSITION Hold or place on the appropriate point.

CHRYSOBERYL

SOURCE Australia, Brazil, Burma, Canada, Ghana, Norway
APPEARANCE Tabular crystals
PROPERTIES A "stone of new beginnings," brings compassion and forgiveness. Increases spirituality and personal power, generosity, creativity, and confidence. Enables seeing both sides. Used with other stones, highlights cause of dis-ease. Balances adrenaline and cholesterol.
POSITION Place as appropriate.

AMAZONITE

SOURCE USA, Canada, Brazil, India, Mozambique, Namibia, Austria
APPEARANCE Opalescent with veins
PROPERTIES Filters information passing through brain and combines with intuition. Elixir is beneficial to all levels of consciousness; balances the metabolism. Heals heart and throat chakra, enhancing loving communication, opens the third eye. Aligns physical and etheric bodies, maintains good health. Balances male–female energy. Beneficial in osteoporosis, tooth decay, calcium deficiency, and calcium deposits. Dissipates energy blocks in nervous system, relieves muscle spasm. Shields from microwaves.
POSITION Hold or place on the appropriate point.

BERYL

SOURCE USA, Russia, Australia, Brazil, Czech Republic, France, Norway
APPEARANCE Prismatic crystals
PROPERTIES Brings a positive view and teaches how to do only what is necessary, filtering distractions and overstimulation. Enhances courage, relieves stress, calms mind. Aids organs of elimination; strengthens pulmonary and circulatory systems, increasing resistance to toxins. Treats liver, heart, stomach, and spine; heals concussion. Elixir treats throat infections.
POSITION Hold or place on the appropriate point.

GREEN CRYSTALS

BLOODSTONE (HELIOTROPE)
SOURCE Australia, Brazil, China, Czech Republic, Russia, India
APPEARANCE Green Quartz flecked with Red Jasper
PROPERTIES An energy cleanser, purifies blood and detoxifies liver, kidneys, and spleen. Benefits all blood-rich organs, regulates blood flow. Cleanses lower chakras, realigning energies. A "stone of courage." Aids recognition that chaos precedes transformation.
POSITION Hold or place on the appropriate point.

GREEN CALCITE
PROPERTIES Rids body of infections. Aids transition from stagnant to positive situation. A mental healer, brings balance.
See Calcite, page 106, for general information.
POSITION Hold or place on the appropriate point.

CHRYSOPRASE
SOURCE USA, Russia, Brazil, Australia, Poland, Tanzania
APPEARANCE Opaque, flecked
PROPERTIES A relaxation stone, encourages peaceful sleep. Promotes hope and eloquence, gives personal insight. Calming and non-egotistical, creates openness to new situations. Draws out talents, stimulating creativity. Energizes heart chakra and body. Enhances fertility, said to guard against sexually transmitted diseases. Aids gout, eye problems, and mental illness; treats heart problems and ameliorates infirmity.
POSITION Place as appropriate.

EMERALD
SOURCE India, Zimbabwe, Tanzania, Brazil, Egypt, Austria
APPEARANCE Bright, transparent gemstone or cloudy crystal
PROPERTIES Gives physical, emotional, and mental equilibrium, focuses intention, raises consciousness, and fosters positive action. Provides inspiration and infinite patience; enhances psychic abilities and wisdom from mental plane. Strengthens memory, aids eloquence, promotes truth, inspires deep inner knowing. "Stone of successful love," bringing domestic bliss and loyalty. Treats lungs, heart, spine, and muscles, soothes the eyes.
POSITION Wear on the little finger, ring finger, over the heart, or on the right arm. Do not wear constantly.

GREEN GARNET (ALMANDINE)
PROPERTIES A healing stone.
See Garnet, page 102, for general information.
POSITION Wear on earlobes, the finger, or over the heart.

GREEN CRYSTALS

JADE (JADEITE, NEPHRITE)

SOURCE USA, China, Italy, Burma, Russia

APPEARANCE Translucent, somewhat soapy feel. Jadeite is translucent; Nephrite has a creamier look

PROPERTIES Calms nervous system, channels passion in constructive ways. Harmonizes dysfunctional relationships. Symbol of purity and serenity, increases love. Releases negative thoughts, soothing mind. A protective stone, brings harmony. "Dream stone," aids emotional release through dreams, manifests dreams. Treats kidneys, removes toxins, rebinds cellular and skeletal systems, helps stitches heal.

POSITION Hold or place on the appropriate point.

GREEN JASPER

PROPERTIES Heals and releases dis-ease and obsession. Useful for skin disorders, dispels bloating.

See Jasper, page 102, for general information.

POSITION Anywhere.

MALACHITE

SOURCE Romania, Zambia, Zaire, Russia

APPEARANCE Concentric light and dark bands

PROPERTIES "Stone of transformation," draws out deep feelings of hurt and resentment, breaks unwanted ties and outworn patterns of behavior. Clears and activates chakras, clarifies emotions, releases negative experiences. Combined with Azurite aids visualization and psychic vision. Aids responsibility for one's actions, thoughts, and feelings. Absorbs plutonium pollution, guards against radiation. Treats asthma, arthritis, fractures, swollen joints, growths, and tumors. Aligns DNA and cellular structure, enhances immune system. Use polished Malachite for elixir preparation.

POSITION Wear on the left hand or place on the third eye.

GREEN AGATE

PROPERTIES Enhances mental and emotional flexibility and improves decision-making. Resolves disputes.

See Agate, page 100, for general information.

POSITION Hold or place on the appropriate point.

PERIDOT (OLIVINE) "CHRYSOLITE"

SOURCE USA, Brazil, Egypt, Ireland, Russia

APPEARANCE Clear crystal

PROPERTIES Cleanser of subtle and physical bodies and mind. Releases negative patterns and vibrations, promotes clarity and well-being. Releases and neutralizes toxins on all levels. Visionary stone, bringing understanding of destiny and purpose, especially helpful to healers. Opens and cleanses heart and solar plexus chakras, regulates cycles of life. Alleviates jealousy and anger, reduces stress, motivates growth and necessary change. Has a tonic effect, healing and strengthening. Used for heart, lungs, spleen, intestinal tract, ulcers; strengthens eyes. Aids birth contractions.

POSITION Wear at the throat.

GREEN CRYSTALS

GREEN AVENTURINE

PROPERTIES Works on heart chakra, activating, cleansing, and protecting. Shields from vampirism of heart energy. Useful healer for body.

See Aventurine, page 103, for general information.

POSITION Over the heart chakra or anywhere.

GREEN SAPPHIRE

PROPERTIES Improves vision, aids dream recall. Stimulates heart chakra, bringing loyalty and fidelity.

See Sapphire, page 119, for general information.

POSITION Touching the body or the finger.

GREEN QUARTZ

PROPERTIES Opens and stabilizes heart chakra. Transmutes negative energy, inspires creativity, balances endocrine system.

See Quartz, page 100, for general information.

POSITION Place on the heart.

TOURMALINE

SOURCE Sri Lanka

APPEARANCE Shiny, opaque; long striated or hexagonal structure

PROPERTIES Cleansing and purifying, transforms dense energy into lighter vibration. Grounds spiritual energy, clears chakras, forms protective shield. Releases tension, helpful in spinal adjustment. Balances male–female energy, right and left brain. Attracts inspiration, compassion, tolerance, and prosperity.

Green Tourmaline opens heart chakra, promotes compassion. One of the most powerful green healing stones, provides balance. Transforms negative to positive energy. Helps problems with father figures. Facilitates study of herbalism, heals plants. Treats eyes, heart, thymus, brain, and immune system; facilitates weight loss; purifies and strengthens nervous system. Relieves chronic fatigue and exhaustion, rejuvenating and inspiring creativity.

POSITION Hold or place on the appropriate point.

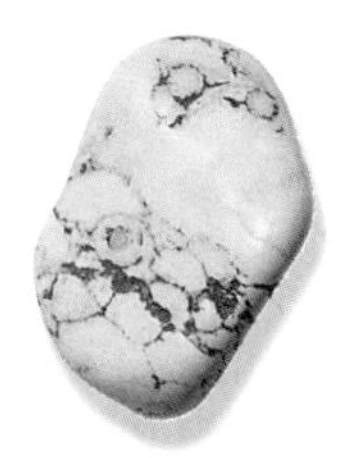

VARISCITE

SOURCE USA, Germany, Austria, Czech Republic, Bolivia

APPEARANCE Opaque, veined

PROPERTIES "Stone of encouragement," useful when dealing with illness and invalids, giving hope and courage. Aids past-life exploration. Heals nervous system; treats abdominal distention, constricted blood flow; regenerates elasticity of veins. Helpful for male impotence.

POSITION Hold or place on the appropriate point.

GREEN CRYSTALS

GREEN APOPHYLLITE

PROPERTIES Activates heart chakra, promotes a forthright heart. Absorbs universal energy. Aids fire walking.

See Apophyllite, page 99, for general information.

POSITION Place on the third eye.

GREEN KUNZITE

PROPERTIES Helps to ground spiritual love.

See Kunzite, page 104, for general information.

POSITION Hold or place on the appropriate point.

GREEN-PINK CRYSTALS

UNAKITE

SOURCE South Africa

APPEARANCE Mottled

PROPERTIES A "stone of vision," balances emotions with spirituality. Facilitates rebirthing, integrates information from past that creates blockages, gently releases conditions inhibiting growth. Reaches root cause of dis-ease. Treats reproductive system, creates weight gain, aids healthy pregnancy.

POSITION Hold or place on the appropriate point.

WATERMELON TOURMALINE

APPEARANCE Pink enfolded in green

PROPERTIES Aids understanding of situations. "Super-activator" of heart chakra, linking it to higher self. Treats emotional dysfunction. Resolves conflict.

See Tourmaline, page 117, for general information.

POSITION Hold or place on the appropriate point.

GREEN AQUAMARINE
See Aquamarine, page 110.

GREEN IRON PYRITE
See Iron Pyrite, page 108.

GREEN CHRYSOCOLLA
See Chrysocolla, page 112.

GREEN TOPAZ
See Topaz, page 97.

GREEN SERPENTINE
See Serpentine, page 124.

GREEN MUSCOVITE
See Muscovite, page 105.

GREEN LEPIDOLITE
See Lepidolite, page 121.

GREEN CAT'S EYE
See Cat's Eye, page 106.

GREEN DIOPTASE
See Dioptase, page 113.

GREEN ONYX
See Onyx, page 124.

GREEN SELENITE
See Selenite, page 99.

ROYAL-INDIGO CRYSTALS

SAPPHIRE

SOURCE Burma, Czech Republic, Brazil, Kenya, India

APPEARANCE Bright, transparent gemstone

PROPERTIES Relaxes, focuses, and calms the mind, releasing unwanted thoughts and mental tension, promoting peace of mind and serenity. Aligns physical, mental, and spiritual planes, restoring balance within body. Supports body to carry out purpose. Releases depression and spiritual confusion, aids concentration. An "eye stone," removing impurities and stress. Brings prosperity, attracting gifts and releasing frustration. Treats disorders of blood, alleviates excessive bleeding. Strengthens veins, improving elasticity.

Royal-Indigo Sapphire eliminates negative energies from chakras, stimulates third eye to access information for growth. Teaches self-responsibility for thoughts and feelings. Treats brain disorders including dyslexia.

POSITION Touching the body or the finger.

AZURITE

SOURCE USA, Australia, Chile, Peru, France, Namibia, Russia

APPEARANCE Very small, shiny crystals (not visible when tumbled)

PROPERTIES Stimulates third eye; facilitates psychic development, spiritual guidance, and higher consciousness. Brings about clear understanding, releasing long-standing blocks in communication and stimulating memory. Brings new perspectives and expands mind. Treats arthritis and joint problems, aligns spine.

POSITION Wear on the right hand or place as appropriate.

LAPIS LAZULI

SOURCE Russia, Afghanistan, Chile, Italy, USA, Egypt

APPEARANCE Opaque, flecked with gold

PROPERTIES A "stone of protection and enlightenment," enhancing dream work and psychic abilities. Quickly releases stress, bringing deep peace. Powerful thought amplifier. Stimulates higher faculties of mind, promoting objectivity and clarity. Promotes creativity and attunement to source. Contacts spirit guardians. Harmonizes physical, emotional, mental, and spiritual levels. Alleviates pain and migraine. Overcomes depression. Benefits respiratory system, throat, cleansing organs, bone marrow, thymus, and immune system. Overcomes hearing loss. Purifies blood, boosts immune system. Alleviates insomnia, vertigo.

POSITION Wear at the throat or place on the forehead.

SODALITE

SOURCE North America, France, Brazil, Greenland, Russia, Burma, Romania

APPEARANCE Mottled dark and light blue-white

PROPERTIES Unites logic and intuition, eliminates mental confusion. Encourages rational thought, objectivity, truth, intuitive perception, verbalization of feelings. Calms mind, allowing new information to be received. Brings about emotional balance. Aids group work. Balances metabolism. Benefits cleansing organs, boosts immune system. Combats radiation and insomnia.

POSITION Hold or place on the appropriate point.

ROYAL-INDIGO CRYSTALS

BLUE JASPER

PROPERTIES Connects to spiritual world. Balances yin and yang energy, stabilizes aura. Sustains energy during fast, heals degenerative diseases. Balances mineral deficiency. Navel and heart chakras for astral travel.

See Jasper, page 102, for general information.

POSITION Hold or place on the appropriate point.

BLUE TOPAZ

PROPERTIES Aids throat chakra and verbalization.

See Topaz, page 97, for general information.

POSITION Wear on the ring finger or position on the throat chakra.

VIOLET CRYSTALS

VIOLET FLUORITE

PROPERTIES Stimulates third eye, imparts common sense to psychic communication. Assists disorders of bones and bone marrow.

See Fluorite, page 106, for general information.

POSITION Hold or place on the appropriate point.

VIOLET AMETHYST

PROPERTIES A lighter vibration of Purple Amethyst with the same properties. This stone works on a profoundly spiritual level.

See Amethyst, page 122, for general information.

POSITION Place on all parts of the body, especially the crown chakra.

BLUE HAWK'S EYE
See Hawk's Eye, page 96.

ELECTRIC BLUE OBSIDIAN
See Obsidian, page 123.

BLUE BERYL
See Beryl, page 114.

BLUE TOURMALINE
See Tourmaline, page 117.

VIOLET CRYSTALS

LAVENDER JADE

PROPERTIES Aids emotional hurt; teaches subtlety and restraint in emotional matters.

See Jade, page 116, for general information.

POSITION Hold or place on the appropriate point.

LEPIDOLITE

SOURCE USA, Czech Republic, Madagascar, Brazil

APPEARANCE Platelike shiny crystals

PROPERTIES Activates throat, heart, third eye, and crown chakras, brings cosmic awareness. Stimulates intellect, reduces stress, relieves despondency, and overcomes insomnia. A "stone of transition," releasing and reorganizing old patterns and inducing change. Restructures DNA. Enhances generation of negative ions. Excellent for menopause. Protects against electromagnetic pollution.

POSITION Hold or place on the appropriate point.

PURPLE CRYSTALS

CHAROITE

SOURCE Russia

APPEARANCE Mottled and veined

PROPERTIES A "stone of transformation." Synthesizes heart and crown chakras, cleansing aura, stimulates unconditional love. Integrates "negative qualities," facilitates acceptance of others. Transmutes energy, converting dis-ease to wellness. Grounds spiritual self. Heals and integrates dualities. Treats eyes, heart, liver, pancreas. Regulates blood pressure. Alleviates aches and pains.

POSITION Place over the heart.

PURPLE FLUORITE

PROPERTIES Stimulates third eye, imparts common sense to psychic communication. Assists disorders of bones and bone marrow. Shields computers.

See Fluorite, page 106, for general information.

POSITION Hold or place on the appropriate point.

VIOLET MUSCOVITE

See Muscovite, page 105.

PURPLE CRYSTALS

PURPLE SAPPHIRE

PROPERTIES A "stone of awakening," aids meditation and rise of kundalini energy, stimulating crown chakra and opening spirituality.

See Sapphire, page 119, for general information.

POSITION Touching the body. Crown.

SUGILITE (LUVULITE)

SOURCE South Africa

APPEARANCE Opaque, lightly banded; or translucent (may be combined with Manganese or Quartz)

PROPERTIES A "love stone," represents perfection of spiritual love, opens chakras to flow of love. Brings spiritual awareness, positive thoughts, promotes channeling ability. Protects soul from shocks and disappointments. Aids forgiveness, eliminates hostility. Excellent for use in autism and learning difficulties. Aids those who do not fit in. Promotes understanding of effect of mind on body. Resolves group difficulties. Benefits cancer sufferers. Place on third eye to alleviate despair. Clears headaches and discomfort.

POSITION As appropriate, especially over the heart.

AMETHYST

SOURCE USA, Canada, Brazil, Mexico, Russia, Sri Lanka, Uruguay, East Africa, Siberia

APPEARANCE Transparent, pointed crystals

PROPERTIES Powerful healer and protector, enhances psychic abilities and spiritual awareness. Stimulates throat chakra, clears aura, transmutes negative energy. Calms mind, enhances meditation and visualization. Disperses psychic attack. Ameliorates anger, rage, fear, and resentment. Aids assimilation of new ideas. Connects cause and effect. Promotes selflessness. Effects enhanced by Rose Quartz. Supports sobriety. Relieves physical and emotional pain. Boosts production of hormones, strengthens cleansing organs and circulatory system, calms nervous system, unites scattered energies. Aids insomnia, hearing, endocrine glands, digestive tract, heart, cellular disorder.

POSITION Place on all parts of the body, especially the throat and chest.

PURPLE TOURMALINE

PROPERTIES Stimulates healing of the heart and produces loving consciousness. Connects the base and the heart chakras, increasing devotion and loving aspiration. A stone of creativity.

See Tourmaline, page 117, for general information.

POSITION Place on the crown.

PURPLE JASPER

See Jasper, page 102.

BLACK CRYSTALS

JET

SOURCE Worldwide, especially USA
APPEARANCE Coal-like
PROPERTIES Draws out negative energy and unreasonable fears. Promotes control of life; fights mood swings and depression. Protects against violence and illness. Stabilizes finances. Increases virility. Treats migraine, epilepsy, swellings, colds.
POSITION Wear anywhere, set in silver.

OBSIDIAN

SOURCE Worldwide
APPEARANCE Volcanic glass, shiny, opaque
PROPERTIES Protective, repels negativity. Transforms negative environmental energies. Disperses unloving thoughts. Very direct stone – brings undesirable attributes to the surface for transmutation. Improves self-control. Aids in letting go of the past and in habit-breaking. Earths base chakra. Retains balance during times of change.
Black Obsidian in particular strengthens prophesy. Used in shamanic ceremonies to remove physical disorder. Do not use for long periods.
POSITION Carry or place as appropriate.

APACHE TEAR

SOURCE USA
APPEARANCE Translucent when held to light, water-worn pebble
PROPERTIES Translucent form of Obsidian. Gentler effect when bringing up negativity. Absorbs negative energies, protects aura. Comforts grief, provides insight into distress. Relieves grievances. Stimulates analytical capabilities and forgiveness. Removes self-limitations.
See also Obsidian, page 123.
POSITION Wear around the lower chakra (men) or heart chakra (women).

BLACK SAPPHIRE

PROPERTIES Most protective sapphire, centers within body and imparts confidence in own intuitions. Excellent for employment prospects and keeping a job.
See Sapphire, page 119, for general information.
POSITION Place touching the body or the finger.

BLACK TOURMALINE

PROPERTIES Deflects and repels negative energy, especially psychic attack. Protects against microwaves, radiation, "spells," and ill-wishing. Grounds spiritual energy, increases physical vitality. Defends against debilitating disease, strengthens immune system. Treats dyslexia and arthritis.
See Tourmaline, page 117, for general information.
POSITION Wear around the neck.

BLACK CRYSTALS

SARDONYX

PROPERTIES A "stone of absorption," clearing negative energies.

See Sardonyx, page 97, for general information.

POSITION Place anywhere.

ONYX

SOURCE Italy, Mexico, USA, Russia, Brazil

APPEARANCE Banded

PROPERTIES Strength-giving, beneficial in difficult or confusing periods. Centers and aligns with higher power. Promotes vigor, steadfastness, stamina, wise decisions. Aids learning lessons, imparts self-confidence, ease in surroundings. Mental tonic, aids fears and worries. Recognizes and integrates dualities. Good for teeth, bones, bone marrow, blood disorders, feet.

POSITION Wear on the left side of the body.

RED-BLACK CRYSTALS

RED-BLACK OBSIDIAN

PROPERTIES Brings about rise of kundalini energy. Promotes vitality, virility and brotherhood. Treats fevers and chills.

See Obsidian, page 123, for general information.

POSITION Hold or place on the appropriate point.

SERPENTINE

SOURCE Britain, Norway, Russia, Zimbabwe

APPEARANCE Mottled, dual color

PROPERTIES Opens new pathways for kundalini energy to rise. Enhances meditation, aids retrieval of ancient wisdom and past lives. A "stone of longevity." Eliminates parasites; aids calcium and magnesium absorption.

POSITION Hold or place on the appropriate point.

BLACK PYROLUSITE

See Pyrolusite, page 125.

BLACK SMOKY QUARTZ

See Smoky Quartz, page 97. Black Smoky Quartz has often been irradiated to produce color (see page 49).

BLACK CALCITE

See Calcite, page 106.

BLACK MUSCOVITE

See Muscovite, page 105.

BLACK-AND-WHITE CRYSTAL

SNOWFLAKE OBSIDIAN

PROPERTIES A "stone of purity," balances body, mind, and spirit. Recognizes and releases wrong thinking and ingrained patterns. Brings about dispassion, inner centering. Treats veins, skeleton. *See Obsidian, page 123, for general information.*
POSITION Hold or place on the appropriate point.

GRAY-SILVER CRYSTALS

CHALCEDONY

SOURCE USA, Austria, Czech Republic, Iceland, Mexico, New Zealand
APPEARANCE Transparent or opaque
PROPERTIES A "stone of brotherhood," promotes group stability, alleviates hostility, promotes benevolence. Increases physical energy. Balances body, emotions, mind, and spirit. Eases self-doubts, facilitates constructive inward reflection. Creates open persona. Absorbs and dissipates negative energies. Cleanses, including open sores. Fosters maternal instinct, increases lactation. Improves mineral assimilation, combats mineral buildup in veins. Aids dementia and senility.
POSITION Wear on the fingers, around the neck, or as a belt buckle. Place as appropriate.

HEMATITE

SOURCE Britain, Italy, Brazil, Sweden, Canada, Switzerland
APPEARANCE Shiny, silver when polished
PROPERTIES Grounding, protecting, and balancing, boosts self-esteem. Aids concentration, focus, willpower, reliability, and confidence. Enhances memory and original thought. Removes self-limitations. A yang stone, balances meridians of body. Dissolves negativity. Beneficial for legal situations. Supports timid women. Strongly affects blood, aids anemia, supports kidney in cleansing blood. Treats leg cramps, nervous disorders, insomnia. Aids spinal alignment, fractures.
POSITION Place at base and top of spine. Hold.

PYROLUSITE

SOURCE USA, Britain, Brazil, India
APPEARANCE On brown matrix, very thin, fan-shaped
PROPERTIES Restructures life, healing disturbances. Gets to the bottom of problems. Changes and stabilizes relationships. Transmutes physical, emotional, and mental bodies. Promotes confidence, optimism, determination. Dispels interference from psychic world. Treats bronchitis, regulates metabolism, strengthens blood vessels, stimulates sexuality.
POSITION Hold or place on the appropriate point.

GRAY CALCITE
See Calcite, page 106.

MAGNETITE (LODESTONE)
See page 96.

GRAY MUSCOVITE
See Muscovite, page 105.

GRAY-SILVER CUPRITE
See Cuprite, page 103.

Index

Resources and Credits

SUPPLIER OF CRYSTAL CLEAR
David Eastoe
6 Behind Berry
Somerton
Somerset, UK
Tel: 01458 274633
email: sacredsites.co.uk/sites/petaltone

UK CRYSTAL SUPPLIER
Earthworks
43 Wessex Trade Centre
Ringwood Road
Poole
Dorset, UK
Tel: 01202 717127
Fax: 01202 717128

UK AND USA GEM REMEDY SUPPLIERS
The Flower Essence Repertoire
The Living Tree
Milland
Liphook
Hants, UK
Tel: 01428 741572
email: flower@atlas.co.uk

Centergees
2007 Northeast 39
Portland OR 97212, USA
Tel: 503-284-6603
(also stockists for Crystal Clear)

Flower Essence Pharmacy
6600 North Highway #1
Little River CA 95456, USA
Tel: 800-343-8693
Fax: 707-937-0441

Alaskan Flower Essence Project
P.O. Box 1369
Homer AL 99603-1369, USA
Tel: 907-235-2188
Fax: 907-235-2777
email: info@alaskanessences.com
(manufacturers of gem elixirs)

The publishers are grateful to the following for permission to reproduce copyright material:
(*l* left, *r* right, *c* center, *t* top, *b* bottom, *bg* background)
Simon Muncer/Earthworks (*4bg, 99b*),
Historic Royal Palaces (*8c*),
Rex Features (*9t, 19bg*),
Tony Stone Images (*10bl, 24l, 43bl, 50l, 65tr, 80cr, 82cr, 89tr*),
Elizabeth Whiting (*11tl*),
Fortean Picture Library (*28*),
Mary Evans (*30b, 44cl*),
Telegraph Colour Library (38tr, *42c, 66tr*),
Bridgeman (*41r*),
Superstock (*43tr, 78c, 84c*),
Eberhard Thiem/Lotos-Film (*54, both images*),
Image Bank (*72cl, 88bl*),
Geoscience Features Picture Library (*100bl, 103c, 105t*)
The kirlian pictures on *page 53* are by Dr Robert Jacobs
Picture researcher Polly High

The crystal wisdom kit on page 61 is courtesy of Eddison Sadd

With thanks to
Francis Annette, G. Applebee,
Neil Bell, R. J. Clarke,
Angela Enahoro, Patricia James,
A. Ferguson, Alison Honey,
Atsuo Murakami, K. Newton,
Rosemary Nobbs,
and to Lucianne Lasselle
for help with photography
and to
Earthworks, Poole, England
Curiouser and Curiouser of Brighton, East Sussex, England;
Winfalcon of Brighton, England;
Kernowcraft of Perranporth, Cornwall, England;
R. Holt and Co. Ltd. of London, England
for the use of props